Also by Kenneth R. Miller

Only a Theory
Finding Darwin's God

THE
HUMAN
INSTINCT

How We Evolved to Have Reason, Consciousness, and Free Will

Kenneth R. Miller

Simon & Schuster

New York   London   Toronto   Sydney   New Delhi

Simon & Schuster
1230 Avenue of the Americas
New York, NY 10020

First Simon & Schuster hardcover edition April 2018

SIMON & SCHUSTER and colophon are registered trademarks
of Simon & Schuster, Inc.

For information about special discounts for bulk purchases,
please contact Simon & Schuster Special Sales at 1-866-506-1949
or business@simonandschuster.com.

The Simon & Schuster Speakers Bureau can bring authors to
your live event. For more information or to book an event, contact
the Simon & Schuster Speakers Bureau at 1-866-248-3049 or visit
our website at www.simonspeakers.com.

Interior design by Lewelin Polanco

Manufactured in the United States of America

10 9 8 7 6 5 4 3 2 1

Library of Congress Cataloging-in-Publication Data

Names: Miller, Kenneth R. (Kenneth Raymond), 1948- author.
Title: The human instinct : how we evolved to have reason, consciousness, and
free will / Kenneth R. Miller.
Description: First Simon & Schuster hardcover edition. | New York :
Simon & Schuster, [2018] | Includes bibliographical references and index.
Identifiers: LCCN 2017006770 (print) | LCCN 2017016386 (ebook) |
ISBN 9781476790282 | ISBN 9781476790268 (hardcover : alk. paper) |
ISBN 9781476790282 (ebook)
Subjects: LCSH: Evolutionary psychology. | Human evolution. | Behavior
evolution. | Consciousness. | Natural selection. | Human behavior.
Classification: LCC BF698.95 (ebook) | LCC BF698.95 .M55 2018 (print) | DDC
155.7—dc23
LC record available at https://lccn.loc.gov/2017006770

ISBN 978-1-4767-9026-8
ISBN 978-1-4767-9028-2 (ebook)

To my teachers, especially the incomparable Mr. Paul Zong, who first led me to discover the wonders of biology.

Contents

Contents

Prologue

S tories matter. And once, we had one.

We knew our place. We were the first fruits of creation, stewards of the Earth, masters of the living world. Whether we traced our kind to a rebellious first couple or the upward climb of the Dineh, our people, into the Fourth World, we had a story. That story, or more properly those many stories, confirmed the dignity and value of human existence. They set us apart from the animals. They assured us that our actions mattered, our choices were real, and our lives fit into a fabric of significance.

To be sure, those visions were not all sweetness and light. Many were filled with darkness, many reflected the depths of the human spirit, and many served to spark savage excesses of passion, greed, and even murder. Yet, even in the worst of ages, those narratives filled one of the most basic human needs. They fashioned a sense of place, mission, and value that set our species, for better or for worse, at the pinnacle of the living world. Our Earth was not only the center of the universe; it was home to the only species in that universe that truly mattered.

And then we lost it. Our stories seemed to vanish, and with them our souls, our place in the heavens, and in many ways, ourselves.

The story of that loss has been told many times, sometimes in the context of the enlightenment, of the scientific revolution, or of the great age of discovery. In retrospect, it was surely more triumph than tragedy. The perplexing movements of planets through space yielded to a mathematics of elegance and precision. The bewildering chemistry of matter was reduced to a table of elements, and the elements themselves to aggregates of simple particles. Electricity and magnetism were united, and new tools fashioned to probe ever deeper into the heart of existence.

But of all these great advances, one stood apart in the way it spoke directly to the human conception of self. It was, of course, the theory of evolution by natural selection. To many, it seemed that Charles Darwin's ideas on the origin of species had drained the lifeblood from our comforting self-portraits. The old certainties were truly gone, and something new had to take their places. But what? In a sense, we had become "Darwin's people," but what could that possibly mean?

Not surprisingly, many were not willing to let the old stories go quietly. While some, like Harvard botanist Asa Gray, were quick to accept Darwin's great idea, others fought back as though civilization itself were at stake. Books were censored, teachers put on trial, and laws passed to prevent students from learning of any theory teaching that "man has descended from a lower order of animals." One such law, in the state of Tennessee, led to the infamous Scopes "monkey trial" in 1925. That law stood until the Supreme Court struck it down in 1968, but even that great court could not strike down popular resistance to so subversive and revolutionary an idea as evolution.

Even today, many fight back by attacking evolution itself, casting themselves as "creationists" who reject broad areas of consensus in modern science. To them, cosmology, astronomy, physics, and even geology have conspired to spin an evolutionary epic that is, as one American politician recently claimed, a "lie from the pit of hell." Others advance an idea known as "intelligent design" in which the mechanisms of evolution are rejected as inadequate to account for the complexity of living things. Instead, the actions of a "designer" are invoked, an intelligent agent standing outside of nature while serving as the grand architect of life. In 2005, this was exactly the argument made in *Kitzmiller v. Dover*, a highly publicized federal trial in Pennsylvania, a trial in which I served as lead witness against "intelligent design."

What both these lines of attack have in common is the call for evolutionary theory to be discarded and replaced by something radically different. The motivations in each case, sometimes expressed quite openly, are not so much to "correct" a scientific error as they are to replace science itself with a view of human origins consistent with certain religious teachings.

As interesting as it might be to take these arguments apart, point by point, that has already been done, not just in the *Kitzmiller* trial[1] but in a host of popular books by scientists and science writers. No point in beating that poor horse again. But I don't think the concerns of all who resist evolution should be dismissed as naïve, trivial, or uninformed. In fact, the passionate unease with which some of evolution's critics regard many of its messages proclaimed in the name of science speaks to the humanist within many scientists, including myself. I believe this unease derives not so much from how we came to be, but rather from what we should make of ourselves as creatures of evolution. In other words, such discontent arises from a fear that accepting the theory of evolution suggests that we are mere *products* of evolution, neither God's people nor Darwin's, but just another of a multitude of creatures pointlessly struggling for existence.

To many, there is a sense that accepting evolution also means accepting a worldview that denies the significance of the human species, explains away our social institutions as artifacts of natural selection, and depicts individual thought and behavior as robotic responses to inputs from the environment. As Sam Harris puts it in his book on free will, "the idea that we, as conscious beings, are deeply responsible for the character in our mental lives and subsequent behavior is simply impossible to map onto reality."[2] However we may regard ourselves, we are driven, according to Harris, by forces over which we have no real control. In Harris's interpretation of the evolutionary narrative, we seem to be nothing more than casual throw-offs, byproducts of a universe far greater than our imagination, a universe in which we are no more than thoughtless works of nature.

Evolution, this line of thinking goes, is driven entirely by natural forces, by principles that apply to living and nonliving matter alike. If, as Stephen Pinker writes, science has exposed "the absence of purpose in laws governing the universe,"[3] then clearly it means that there is an "absence of purpose" in the evolutionary process itself. In our modern, sophisticated, rational world, those who hold this view of evolution regard the human presence as nothing special. They see us as cosmic accidents of no significance, depict human art and creativity as the pointless byproducts of natural selection, and regard purpose, self, and even consciousness as chemical illusions that signify nothing, whatever their sound and fury. They, in short, grimly accept the view that we humans matter very little in the grand scheme of things. The story of human evolution, according to those who spin this narrative, is one of pointless accident, dark struggle, and ultimate meaninglessness. No wonder so few want to hear the bad news.

But there is something both illogical and unsound with any narrative that depicts a species able to unravel the story of evolution as insignificant carbon-based fuzz on the surface of a small blue planet. In fact, I emphatically believe there is something special about *Homo sapiens*, something that truly sets us apart. So it is imperative to ask if we need a fundamental revision of evolutionary theory to account for the specialness of human nature. As we will see, I don't think so. What we really need is to understand and appreciate the beauty and subtlety of evolution in greater depth than ever.

We are living creatures, to be sure, one species among countless millions that have come and gone in our planet's lifetime. But we are also uniquely the creatures of music and art, of poetry and laughter, of science, reason, and mathematics. We are the children of evolution in every sense, but we are children of the universe as well, and from that realization comes a new and exhilarating way to see our place among other living things and our home among the stars. It is exactly that place I propose to explore in the pages that follow.

Chapter 1

Grandeur

I think Charles Darwin might have seen his critics coming. Unlike most nineteenth-century works of science, *On the Origin of Species* is still read today. Much of that attention has been earned by the logical power and simplicity of Darwin's argument. He begins with a chapter on variation among domestic animals and plants, something that every animal and plant breeder in the England of his time would have been familiar with. Chapter 2 points out that similar variation exists in wild species. Having established that individual members of a species vary in their characteristics, chapter 3 then describes a "struggle for existence" occurring everywhere in the natural world, producing forces that work remarkably like the hand of a breeder to shape the characteristics of every living species. At that point, the stage is set for the theory of evolution by natural selection, which he introduced by name in chapter 4. The remaining ten chapters enlarge and expand upon the evidence for this theory. The book has been called "one long argument," and so it is. A powerful and elegant argument.

But there is another reason *The Origin* is not only read today, but also widely quoted. While much of the book is mired in scientific minutiae and arcane speculation, as it moves toward a conclusion, *The Origin* shines with a clarity—even a kind of poetry—rarely seen in a scientific document. In particular, having brought his many arguments to their logical conclusions, Darwin seems compelled to tell us what a wonderful vision of nature he has set before us:

> When I view all beings not as special creations, but as the lineal descendants of some few beings which lived long before the first bed of the Silurian system was deposited, they seem to me to become ennobled.[1]

And why are they "ennobled"? To Darwin, it is because living species are linked to an almost endless history of struggle and success, often against great odds. So distant is that past, so persistent are the triumphs of those shaped by natural selection, that we may look at them with pride, confident of an equally long and glorious future.

> As all the living forms of life are the lineal descendants of those which lived long before the Silurian epoch, we may feel certain that the ordinary succession by generation has never once been broken, and that no cataclysm has desolated the whole world. Hence, we may look with some confidence to a secure future of equally inappreciable length. And as natural selection works solely by and for the good of each being, all corporeal and mental endowments will tend to progress towards perfection.[2]

Every day, in every way, they're getting better and better—and so are we. The future is secure, and we're getting closer to perfection. Fine words, even though most biologists today, myself included, would argue that evolution never produces "perfection." In fact, it never even gets close. Success in the struggle for existence is all that matters, so being just good enough to get by is good enough. Always has been, always will be. But Darwin spun things differently.

As stirring as these words about perfection may have seemed to nineteenth-century readers, the final paragraph of *The Origin* reaches even higher. Darwin wants us to find beauty in the apparent chaos of nature, using the metaphor of a tangled bank alongside a stream to represent the creativity of the evolutionary process:

It is interesting to contemplate an entangled bank, clothed with many plants of many kinds, with birds singing on the bushes, with various insects flitting about, and with worms crawling through the damp earth, and to reflect that these elaborately constructed forms, so different from each other, and dependent on each other in so complex a manner, have all been produced by laws acting around us.[3]

And finally, just in case his readers might be a bit distressed by the realizations that they are merely the products of "laws acting around us," he assures us that there is indeed something special, something glorious about the whole process:

There is grandeur in this view of life, with its several powers, having been originally breathed into a few forms or into one; and that, whilst this planet has gone cycling on according to the fixed law of gravity, from so simple a beginning endless forms most beautiful and most wonderful have been, and are being, evolved.[4]

It's a stirring sentence. I have often quoted it in my own writings and lectures, and I'm not alone. But if his ideas were on such firm footing, as they clearly were, why did Darwin find it necessary to describe his vision as one of "grandeur"? I think it may have been because he recognized full well that many, if not most, of his readers would surely think otherwise. If we find our origin in the natural world by means of natural laws, then how can we possibly consider humankind as something apart from the beasts of the field, or even the slimy critters of the soil? *Punch*, the humor magazine, picked up on this much later with a satirical cartoon on its cover, stating "Man is but a worm."[5] Building on Darwin's own writings, the cartoon depicted an earthworm-like creature first arising out of chaos, then morphing into a series of monkeys, next a cave man, then an English aristocrat, and finally into Darwin himself. Hardly a vision rooted in grandeur.

Darwin clearly realized that a little polishing of the human ego would go a long way toward encouraging acceptance of his ideas, and that is exactly what we see in the concluding paragraphs of *The Origin*.

He understood that most would not find this vision "grand" and decided to do what he could to convince them otherwise. But I'm not sure this appeal to his readers to recognize the "grandeur" of evolution ever took hold. And I believe that remains the case today, even among many who fully accept the evolutionary story of our origins.

In Ian McEwan's novel *Saturday*, his contemporary protagonist begins the single day of the story's title by contemplating Darwin's use of that very word. As Henry Perowne, a London neurosurgeon, rises, the phrase comes to him over and over again: *There is grandeur in this view of life.* Three times he repeats those words, and then remembers why. Last night, in the bath after a tiring day, he had skimmed a biography of Darwin sent him by his "all too literate" poet daughter, Daisy. He doesn't remember much—he'd never actually read Darwin himself—but that phrase stuck with him. Musing to himself, he contemplates the forces that drove the great naturalist to compose the final sentence of his masterwork:

> Kindly, driven, infirm Charles in all his humility, bringing on the earthworms and the planetary cycles to assist him with a farewell bow. To soften the message, he also summoned up a Creator in later editions, but his heart was never really in it. Those five hundred pages deserved only one conclusion: endless and beautiful forms of life, such as you see in a common hedgerow, including exalted beings like ourselves, arose from physical laws, from war of nature, famine, and death. This is the grandeur. And a bracing kind of consolation in the brief privilege of consciousness.[6]

We emerge from war, famine, and death, and all we have to show for it is the "brief privilege of consciousness"? Having rushed headlong through his medical studies and into practice, Perowne, who describes himself as not having touched a non-medical book for fifteen years, permits himself a brief contemplation of the meaning of Darwin's work. Although a nonbeliever, it leads him to think of religion. He recalls the words of Philip Larkin, where the poet wrote that if he ever needed to "construct a religion," he would make use of water.

Perowne, the rationalist, doesn't hold much stock in Larkin's answer.

But he thinks to himself that if he were ever "called in" to construct a religion, instead of water,

> . . . he'd make use of evolution. What better creation myth? An unimaginable sweep of time, numberless generations spawning by infinitesimal steps complex living beauty out of inert matter, driven on by the blind furies of random mutation, natural selection and environmental change, with the tragedy of forms continually dying, and lately the wonder of minds emerging and with them morality, love, art, cities—and the unprecedented bonus of this story happening to be demonstrably true.[7]

Demonstrably true, but nonetheless often uninspiring. As he goes about his business, Perowne watches a massive demonstration against the invasion of Iraq but is strangely detached from it by his willingness to appreciate arguments on both sides of the debate. The same profound rationality leads him to dismiss "magical realism" in literature, even though his daughter urges him otherwise. As the day wears on, a minor traffic accident followed by an attempt by the other driver at extortion places Henry and ultimately his family in danger.

In what might fairly be called the climax of the novel, Henry's apartment is invaded and his family is held at knifepoint by Baxter, the extortionist. Perowne's daughter is forced to strip naked, at which point Baxter notices a book of poems with the name "Daisy Perowne" inscribed on the cover. Intrigued, he demands she read one of her poems. As she seems to comply, Baxter is so taken by the beauty of the poem that he asks her to read it again—at which point it becomes clear to her father and the reader that Daisy isn't reading one of her own poems at all. Instead, she's recited, from memory, Matthew Arnold's classic "Dover Beach." For Baxter, the second reading is mesmerizing. His mind seems to wander, which leads to a distracted confrontation in which Perowne and his son are able to overpower and disable the intruders. Afterward, the family realizes that Daisy's choice of Arnold's poem, which she had memorized in her youth, had been their literal salvation.

McEwan, the author, clearly wanted his readers to contemplate the particular poem Daisy recited to Baxter. As if to emphasize this point, he included the full text of "Dover Beach" on two pages following the conclusion of the novel. It makes a fitting afterword to a novel that began with ironic references to Darwin's view of the grandeur of life. The poem's thirty-seven lines contain a deeply thoughtful and melancholy reflection on the onrush of the modern age in mid-nineteenth-century Britain. As Arnold writes:

> The Sea of Faith
> Was once, too, at the full, and round earth's shore
> Lay like the fields of a bright girdle furled.
> But now I only hear
> Its melancholy, long, withdrawing roar,
> Retreating, to the breath
> Of the night-wind, down the vast edges drear
> And naked shingles of the world.

To Arnold, the world has changed, changed utterly. The roar of the ocean at Dover seems only to "bring the eternal note of sadness in," and the modern age, with all its wonders and delights, "Hath really neither joy, nor love, nor light, nor certitude, nor peace, nor help for pain." The same is surely true of the day that Henry Perowne, the successful neurosurgeon, has just endured. The disruptions of the modern age, as described in *Saturday*, intrude despite one's best efforts to find certitude, joy, and peace. And that promise of grandeur seems to fade away as surely as the ebbing waves at Dover Beach.

A SCIENCE OF LIFE

Arnold's poem mirrors much of the popular reaction to Darwin. Once the "sea of faith" was full and round the Earth. But today we see only its "long, withdrawing roar" as evolution displaces the old certainties. Arnold penned "Dover Beach" before *The Origin*, but he published it in 1867, well after Darwin's book had shocked much of Victorian society. Ever since, it has been seen as emblematic of the crisis of faith brought about by the emergence of modern science. And, as McEwan's novel demonstrates, that crisis has not abated.

Setting aside, at least for a moment, the sentiments of artistic intellectuals such as Arnold and McEwan, it's only fair to ask whether and how such concerns have affected the larger culture. In the United States, where outright rejection of evolution is common, one might ask how this came to be. Ironically, one could make a strong argument that it was our country's enlightened drive for universal high school education that brought it on.

Although the United States helped to pioneer free public education, the level of such schooling did not usually extend to the secondary level until the beginning of the twentieth century. Indeed, only one of my four grandparents, all born around the turn of that century, was educated past the eighth grade. But as states began to mandate higher levels of education, schools expanded and with them the demand for teachers and instructional material such as textbooks. As historian Adam R. Shapiro explains in his book, *Trying Biology*,[8] this led New York–based textbook publishers to expand their offerings beyond the basic lessons in botany and zoology that had been part of the curriculum up to that time. Specifically, they offered new books geared to biology itself as a secondary-level discipline. These texts were skillfully marketed by local and regional sales agents, and in line with the social optimism of the times, had a distinct focus on applying scientific knowledge for the betterment of society. The title of one such text, George Hunter's *Civic Biology*, reflected this trend and drew broad conclusions as to how evolutionary principles might be applied to improve society. As such, the book discussed personal hygiene, proper social behavior, and even eugenics. This, of course, was the very textbook used by substitute teacher John Scopes in Dayton, Tennessee.

Compulsory high school education appeared first mostly in urban school districts. This led to a concern that many of the instructional materials clashed with the more rural values of states such as Tennessee, where evolution was regarded as just such an "urban" value. Also, as Shapiro notes, in many states, interactions between local school districts and avaricious publishers persuaded state authorities to wrest control of textbook purchases from individual schools. This led to state oversight of instructional materials and opened the door to legislative battles over the content of textbooks, battles that persist to the present day. It was in this context that the State of Tennessee passed a law, early in 1925, leading directly to the trial of that substitute biology teacher just a few months later.

The Scopes "Monkey Trial," held in Dayton, Tennessee, in 1925, is widely regarded as one of the key events in the social history of the United States. To many Americans, the Scopes trial represents a heroic battle in which reason and science were pitted against ignorance and superstition. The trial was loosely dramatized in the 1955 play *Inherit the Wind*, which has been adapted for television and motion pictures no less than four times. Evolution, of course, serves as the stand-in for enlightenment and reason in that battle. One of the authors of the play, Jerome Lawrence, made this explicit, admitting in an interview that "we used the teaching of evolution as a parable, a metaphor for any kind of mind control [. . .] It's not about science versus religion. It's about the right to think."[9] In the context of the 1950s, when the play first appeared, that lesson might have been applied to the McCarthy hearings. In more recent revivals, however, it is often seen as a statement about the political power of the religious right in America.

But there is an important aspect to the actual Scopes trial that is often overlooked in the rush to draw contemporary lessons from its history. The Butler Act, the Tennessee statute under which John Scopes was prosecuted, did not actually forbid the teaching of evolution, despite a preamble proclaiming its intent to "prohibit the teaching of the Evolution Theory." Instead, the act merely made it unlawful to "teach any theory that denies the story of the Divine Creation of man as taught in the Bible, and to teach instead that man has descended from a lower order of animals." In other words, it was perfectly okay to teach the evolutionary process as applied to oak trees, spider monkeys, whales, and dinosaurs. But leave *Homo sapiens* out of it!

Incredibly, under the Butler Act, one could have taught Darwin's *On the Origin of Species* cover to cover, since that great work actually said nothing about the origin or descent of man. As Darwin scholars know, his thoughts on those issues would come nearly a decade after *The Origin*. John Scopes, of course, was convicted of violating the Butler Act, and although his conviction was set aside on a technicality,[10] the law remained in force until 1967.

Significantly, the language of the Butler Act was typical of antievolution legislation in many states, including the Arkansas statute invalidated in a landmark 1968 Supreme Court case (*Epperson v. Arkansas*). That law, passed by popular referendum forty years earlier, also focused on the question of human evolution, making it unlawful for any instructor to teach "the doctrine or theory that mankind ascended or descended from a lower order of animal." In retrospect, one might wonder why these statutes were worded in this very precise way, to single out human evolution rather than Darwinian evolution in general. After all, if the history of life on our planet is characterized and explained by evolution, doesn't that mean our own history is as well?

Organized antievolution groups appreciate exactly this point, and for that reason they strongly oppose just about anything in mainstream science that is consistent with the natural history narrative of evolution. That means disputing the big bang, the age of the Earth, the geologic ages, the abiotic origin of life, and especially the notion that the fossil record contains any evidence of speciation or change over time. They recognize, quite logically, that if science can demonstrate the evolution of *anything*, then their whole project of depicting humanity as a unique and special creation is doomed.

Most people, however, look at things a bit differently, and the focus of their attention is indeed squarely on the human animal. A recent poll[11] in the great state of Texas, well and justly known as a hotbed of antievolution sentiment, demonstrates this. When Texas voters were asked whether life had existed in its present form since the beginning of time, just 22 percent agreed. In contrast, 68 percent asserted that life had "evolved over time." That might seem to be a stunning result in such a state, but two elements of this particular question were clearly responsible for the 3:1 margin in favor of evolution. First, the question did not mention human evolution. Second, and just as important, one of the possible answers, which garnered 53 percent support among the respondents, stated life had "Evolved over time, entirely through 'natural selection,' but with a guiding hand from God."[12] By keeping any reference to humans off the table, and by including a response that allowed people to choose evolution without seeming to reject their faith, a large majority of Texans supported evolution by natural selection.

What happened when the same polling group was asked about human evolution? Suddenly the numbers changed. Even when presented with a "God guided the process" explanation, only 50 percent agreed that humans evolved over time, while fully 38 percent asserted that "God created human beings pretty much in their present form about 10,000 years ago." When an even more direct question was asked, support for evolution turned into outright rejection. Did human beings as we know them develop from earlier species of animals? Now just 35 percent agreed, while 51 percent disagreed.

It is true, of course, that nearly all of this resistance is religiously inspired. So, a simplistic analysis of the "problem" might suggest that in the absence of religion, acceptance of evolution would rise to the high levels we see in secular European cultures today. But that assumes that mere acceptance, however grudging, is a goal to be sought, and that secular cultures have a better understanding of what it means to be human. I'm not sure that is true. Still, among many who embrace Darwin's legacy, there remains a pessimism, a deep restlessness regarding its ultimate message. To these folks, evolution subverts the once profound distinction between beast and human, it tells us we do not stand at the pinnacle of the living world, and it bequeaths a legacy not from the gods or the stars, but rather one written by the grim dictates of survival, chance, and reproduction. In this view, there may be truth in evolution, but it seems to be a truth that drags us into the muck of struggle and strife rather than lifting us to the imagined heavens of our noblest selves. To the fictional Henry Perowne, this may have been just one more part of the mundane reality of contemporary life, but it is hardly something new. It has, in fact, been part of the heavy baggage of evolutionary thought from its very beginnings, articulated by one of the founders of the theory itself.

DOUBTS OF A FATHER

Nearly all creation stories agree on one thing, which is the uniqueness of the human species and the need for a special story to explain how we came to be. At a fundamental level, the idea of evolution undermines these stories, whether they are set in an Abrahamic Eden or upon the sacred mesas of the Hopi. By telling us that we do not have such a story, by placing our origins squarely in the ordinary genetic, environmental, and selective processes that have produced every other living thing, evolution sweeps such narratives away and leaves us searching for our birthright as thoughtful, intelligent, and hopeful creatures. One of those troubled by that search was in fact a founding father of the theory of evolution itself, Alfred Russel Wallace.

As students of biology learn, Wallace shares full credit with Darwin for the theory of evolution by natural selection. It was in fact a letter from Wallace that moved Darwin to publish his long-held views on natural selection, resulting in papers by both naturalists in 1858, and then in the publication of Darwin's *On the Origin of Species* a year later. A tireless defender of the importance of natural selection, Wallace actually preceded Darwin in proposing that our own species had its origin in the evolutionary process. His 1864 paper, "The Origin of the Human Races and the Antiquity of Man Deduced from the Theory of Natural Selection," traced the physical evolution of the human body to the very same evolutionary forces that had shaped other species. However, Wallace also insisted that as human culture developed, it changed the rules of the evolutionary game. Michael Shermer describes Wallace's thoughts about this in his biography of the naturalist[13]: "Once the brain reached a certain level, however, natural selection would no longer operate on the body because man could now manipulate his environment."

By itself, this was hardly a controversial assertion. But a few years later, Wallace went further, insisting that certain uniquely human attributes could not have been produced by natural selection. Noting that even the "lowest races" possessed the mental attributes necessary to practice the high cultural arts and sciences characteristic of European civilization, Wallace wondered how natural selection could have produced these qualities when they did not seem to be useful to those "in the very lowest state of civilization." He wrote that unless Darwin could show him how talents such as sophisticated musical skill could have aided survival in the struggle for life, "I must believe that some other power [than natural selection] caused that development."[14]

Putting it bluntly, Wallace wrote, "How then was an organ developed far beyond the needs of its possessor? Natural selection could only have endowed the savage with a brain a little superior to that of an ape, whereas he actually possesses one but little inferior to that of the average members of our learned societies."[15] Later in life, Wallace was involved in spiritualism and any number of scientifically questionable pursuits, but as Shermer points out, his argument here was based on none of these. Rather, it hinged "on the failure of natural selection to account for a *variety* of features" critical to human nature itself. [16]

That deep desire to look into the mirror of human nature and find something special still exists. But doesn't evolution devalue that claim? Doesn't it state that the qualities we so treasure, from language to artistic creativity to our high-minded moral codes, arise from nothing more than the grim calculus of competition and survival? When I confront skeptical audiences on the issue of evolution, I find very few individuals genuinely passionate about things such as the reptile-to-mammal transition or the evolution of the vertebrate body plan. What bugs a large number of folks to the core, however, is the idea of *human* evolution. The notion that we crawled out of the slime, that our ancestors were "monkeys," and that our senses of beauty, love, and morality were carved from nature red in tooth and claw seem to them profoundly degrading and demeaning.

Even some present-day scientists, such as Francis Collins, who has headed both the Human Genome Project and the National Institutes of Health, worry about the same issues that troubled Wallace. Collins describes a universal "moral law," a grasp of the concepts of right and wrong that is found in all people, regardless of their specific cultures. On the basis of evolution, he believes that one cannot account for either this moral law or the self-sacrificing altruism that so many people exhibit daily. Therefore, Collins, very much like Wallace, believes that only a higher power could have placed these noble standards of behavior within us.[17]

To many people, as to Wallace and Collins, the idea of accepting human evolution is more troubling than the mere abnegation of a biblical myth. It is a blow to their fundamental sense of what it means to be human. The problem for Wallace and Collins is not that evolution is wrong so much as that it fails to supply a complete and satisfying explanation of what it means to be human. To them and many others, the raw and simple forces that so clearly drive evolution by natural selection do not seem to explain the depth and complexity of human life and thought. Something else is needed.

A CHILLING DOCTRINE?

The view that evolution threatens humanity's traditional view of itself is widely shared. Writing in the *Boston Review*, psychologist Tania Lombrozo put the problem this way:

> People find it dehumanizing to conceptualize themselves as animals, and human evolution underscores the continuity between humans and our (distant) cockroach cousins. . . . Associating animal characteristics with humans has been used to justify inhumane treatment; it strips people of human uniqueness and certain aspects of agency and moral consideration. An evolutionary history shared with other animals—and even plants and bacteria—might threaten the separation between human and non-human that maintaining our "specialness" seems to require.[18]

Dr. Lombrozo goes further, describing a study in which college undergraduates were asked how they thought accepting evolution as true might affect individuals and society.[19] Given the fact that the students in the study held a wide range of views, from fully accepting of evolution to fully rejecting, one might expect that pro-evolution individuals would see acceptance as a positive development, while creationist students might regard it as negative. Surprisingly, that was not true. Students across the board "viewed the consequences of accepting evolutionary principles in a way that might be considered undesirable: increased selfishness and racism, decreased spirituality, and a decreased sense of purpose and self-determination." For example, fully 83 percent of both creationist and evolutionist groups thought the theory would increase selfishness. Similarly, both groups agreed that evolution lessened one's sense of purpose, and that it would tend to increase racist feelings among those who accepted the theory as valid.

As this study shows, the notion that the idea of evolution is destructive to the social fabric is not limited to those who reject the theory for religious reasons. One such person is the celebrated novelist and essayist Marilynne Robinson. Author of books such as *Lila*, *Housekeeping*, and *Gilead*, for which she received the Pulitzer Prize, Robinson has expressed deep unease with the implications of evolution for Western culture and society. This concern was addressed in "Darwinism," the key piece in her 1998 collection, *The Death of Adam: Essays on Modern Thought*. While clearly a critic of what she calls "Darwinism," Robinson is not interested in a scientific attack on evolution itself. She characterizes Darwin's work as "impressive," and notes that evolution, "the change that occurs in organisms over time,"[20] "was observed and even noted in antiquity."

But Robinson is deeply troubled by what she regards as the baggage that Darwinian theory has accumulated in the name of science. After exempting the "phenomenon" of evolution itself from her criticism, she defines "Darwinism" as "the interpretation of the phenomenon which claims to refute religion and to imply a personal and social ethic which is, not coincidentally, antithetical to the assumptions imposed and authorized by Judeo-Christianity."[21] Those "assumptions," as Robinson makes clear, include the bedrock foundations of Western culture regarding the worth of the individual and even the intellectual sources of science itself.

True or not, Robinson makes it clear that she regards evolution, with its emphasis on competition and survival, as a "chilling doctrine." She links it to the extermination of native peoples, to a harsh disregard for the value of the individual, and above all to a bitter reduction in the value of human life, thought, and creativity. Robinson's title, *The Death of Adam*, speaks, of course, to the way in which the idea of human evolution has displaced the Abrahamic creation story of Genesis and the Fall that once accounted for the origins of our species. But the loss, she feels, extends far beyond biblical myth to the very core of humane values and human culture. Quoting from Robinson's book, one reviewer captured her concerns this way:

> The question, as Robinson puts it, is whether "all that has happened on this planet is the fortuitous colonization of a damp stone by a chemical phenomenon we have called 'life.'" Or, in the words

of an eminent sociobiologist, "an organism is only DNA's way of making more DNA." Think of Plato, Bach, Newton, Rembrandt, Shakespeare; then consider the implications of that "only."[22]

To Robinson, if we are *only* the vessels of our DNA, *only* the products of a mindless struggle for existence, and *only* the fortuitous colonizers of sea and soil, then every scrap of art and music and culture and even science is utterly without meaning or value. As Robinson writes, "It is a thing that bears reflecting upon, how much was destroyed, when modern thought declared the death of Adam."[23]

For a biologist, it might be easy to set aside all such concerns by saying something like "You evolved, so deal with it." And if the question of human ancestry and the natural history of our species were all that was at stake, I might go with that curt dismissal. But some of the most visible public champions of evolution have traveled much further and increasingly propose a view that they say is supported by science in its purest form: that human nature is nothing more than the accidental combination of atoms and their aggregation into molecular assemblies that produce within us illusions of value, purpose, and meaning. As Richard Dawkins has famously written, "The universe we observe has precisely the properties we should expect if there is, at bottom, no design, no purpose, no evil and no good, nothing but blind, pitiless indifference."[24]

Dawkins's view of the universe is markedly at odds with a conviction that has united human cultures from their very beginnings. That is that our very existence is a matter of significance. From such thoughts emerge the creation stories that bind societies together, as well as their collective art, music, literature, and even their science. The drive to understand, after all, comes only partly from a hope that scientific knowledge will be of practical importance. Just as critical is the simple desire *to know*, followed by the satisfaction and joy produced by understanding and born of the conviction that human understanding is our goal and even our destiny.

Does human evolution support this ennobling view? In the minds of many, it doesn't. We may regard our place in the animal kingdom as exalted, but to a biologist we primates can be seen as just one tiny branch in an overgrown forest. In historical terms, we appeared only recently—almost an afterthought on planet Earth—and it would be foolhardy to view the whole of natural history as a process with the purpose of bringing our species into existence.

As astronomer Neil deGrasse Tyson explains, "if the purpose of the universe was to create humans, the Cosmos was embarrassingly inefficient about it. And if a further purpose of the universe was to create a fertile cradle for life, then our cosmic environment has got an odd way of showing it. Life on earth, during more than 3½ billion years of existence, has been persistently assaulted by natural sources of mayhem, death, and destruction. Ecological devastation exacted by volcanoes, earthquakes, and climate change, tsunamis, storms, and especially killer asteroids have left extinct 99.99 percent of all species that have ever lived here."[25]

Looked at this way, the conditions of human evolution reduce us to the status of mere organism, just one among many on this improbable planet. The field of evolutionary psychology may explain why we *think* we're important—such illusions have survival value—but evolution itself says we're not. As Stephen Jay Gould wrote, neither we nor one of our most cherished properties was a sure thing in this cold, harsh world:

> Humans are not the end result of predictable evolutionary progress, but rather a fortuitous cosmic afterthought, a tiny little twig on the enormously arborescent bush of life, which if replanted from seed, would almost surely not grow this twig again or perhaps any twig with any property that we would care to call consciousness.[26]

One is left to suspect that for Gould, even our best attempts to find grandeur in life are merely the illusions of that cosmic afterthought.

MORE BAD NEWS

As Gould suggests, the bush of life does not seem in any way to have been programmed to produce us. The evolutionary process is not predictable, and therefore we are, in every sense, an accidental species. This is exactly the title of a book by Henry Gee, a British paleontologist and evolutionary biologist. In *The Accidental Species*, Gee writes, "There is nothing special about being human, any more than there is anything special about being a guinea pig or a geranium."[27] In fact, if the story of science were written by other organisms, rather than humans, Gee knows that they would see things differently:

> Giraffe scientists would no doubt write of evolutionary progress in terms of lengthening necks, rather than larger brains or toolmaking skill. So much for human superiority. If that's not ignominy enough, bacterial scientists would no doubt ignore humans completely except as convenient habitats, the passive scenery against which the bacterial drama is cast. Now, ask yourself—which of these stories is any more valid than any other, at least as a narrative?[28]

Describing where the book will take his readers, Gee points out one of its key themes:

> I take a brief tour of several attributes that at some time or another have been regarded as unique to humans. These include bipedality, technology, intelligence, language, and finally sentience or self-awareness. It turns out that most if not all have been seen in one or more nonhuman species—or once one has accounted for a human bias in investigating such attributes, they turn out to be no more special than any other feature of any other organism.[29]

Apparently, we have to come to grips with the fact that to some extent at least, all of the human properties we hold so dear are found in other animals. Neither language nor technology nor self-awareness is uniquely human. Therefore, we are indeed no more special than any other organism.

Why do we think and act the way we do? Surely it is because our big brains give us access to a range of data and sensory experiences that no other animal can approach. We can then weigh such inputs and determine our own actions, relying on reason, common sense, and personal preference, right? Well, maybe not. If behavioral traits and tendencies are genetic in nature, meaning they can be inherited, then such traits are subject to natural selection just as surely as any other. At a basic level, this means evolution has shaped our brains, the tools with which we think and experience the outside world.

There may be nothing remarkable about that, until the tools of evolutionary psychology are actually employed to explain specific human behaviors. Only a few years ago, one of those studies hit very close to home for me. Like many men I know, I regard my children as among the greatest treasures in my life. While there are many counterexamples, of course, I am privileged to know many men who are and have been exemplary fathers. These men care for and nurture their children, as well as doing the unglamorous work of changing diapers, preparing meals, and cleaning house. Most important, they are steady and supportive influences on the lives of their children, preparing them for happy and productive lives as they grow. If you asked any of them about their lives as parents, they would tell you, almost certainly, that they made a deliberate and thoughtful choice to be involved in the lives of their kids. Some of them would say they did this for themselves, others for the best interests of the children, and still others would say that it was best for their spouses and their relationship. But all would agree, they made the choice.

Not very long ago, such men were greeted by a headline telling them that all of this was just an illusion. The choice wasn't theirs to make, because evolution had made it for them. The real reason they'd immersed themselves in the care of their children was the size of their testicles. They were just a little too small. Guys with the big ones don't hang out with their kids. They're too busy chasing other women and trying to spread their seed around. And it's because their genes made them do it.

This was the popular interpretation of a 2013 study of testicle size and parenting, published in one of the world's most prestigious scientific journals,[30] and widely discussed in the press. As I read the paper and looked over the data, it was clear to me that one could make a case that the "small testicles = good dads" argument was a little shaky on purely statistical grounds. But that's not the point. The researchers were doing their best to correlate a physiological property with a behavioral one, and there was at least some evidence to suggest they had found something interesting. What stood out was the way in which they had *explained* the source of that correlation, in other words, why it existed in the first place. Specifically, they claimed to have an evolutionary reason to expect exactly such a relationship between gonad mass and child care.

Noting that "evolution optimizes the allocation of resources toward either mating or parenting so as to maximize fitness," they had set out to find whether "human anatomy and brain function reflect a trade-off between mating and parenting investment." And so they did. In this particular case, you might say that evolutionary psychology had found the "real" reasons some men are good fathers—as well as why others repeatedly neglect their families to chase other women. I can almost imagine an unfaithful husband saying, "Honey, I couldn't help it. It's those damn testicles." And, wincing in painful anticipation, I can certainly imagine my wife's response if she ever heard those words from me. Curiously, the researchers seem not to have felt the need to conduct an actual survey of the marital fidelity of the men in their study to see if it matched their expectations. Perhaps they felt that their story of evolutionary imperatives was just so compelling that further testing was not required.

In one sense, evolutionary psychology is a straightforward science that seeks to uncover some of the powerful forces that shape human behavior. It often uses empirical measures (testicle mass is an obvious one, effective parenting somewhat less so) to construct explanations based on evolutionary logic (maximization of long-term reproductive fitness). In so doing, it promises to provide valuable insights into basic questions in behavioral science ranging from the personal to the social. One practitioner of the discipline has even suggested using the findings of evolutionary psychology to redesign the social structure of an American city along sound evolutionary principles.[31] But in another sense, evolutionary psychology suggests that our most intimate thoughts, our goals, our values, even our morals are not our own, but are the artifacts of thousands of generations of natural selection, exerting a power that is beyond our ability to control.

Explanations abound in the literature purporting to explain intelligence, racism, sexual orientation, morality, and religious faith all in terms of evolutionary advantage. At a conference in 2009,[32] E. O. Wilson, the author of *Sociobiology*, a founding text of the field, showed slides of the beautiful landscaping surrounding the corporate headquarters of the John Deere corporation in Moline, Illinois. Wilson wondered, why do we find great lawns clustered with shrubbery and lakes so appealing? His answer was that they resemble the environments of the Pleistocene in which evolution formed our species, shaping its behavior, its likes and dislikes, and even its aesthetic tastes. Apparently, our prehistoric ancestors were really into rolling lawns, manicured gardens, and dancing fountains.

If that smacks just a bit of overreach, imagine a future in which evolutionary science will produce definitive answers as to why we prefer Mozart to Salieri, why we regard pedophilia with disdain, and why human societies tend to place mostly males in positions of leadership. That is exactly the program advanced by Wilson in *Consilience*, a landmark 1998 book that made such promises for the future of the evolutionary project. That book, in effect, told my colleagues in the humanities and social sciences to step out of the way, because evolutionary psychologists were taking over their disciplines. And perhaps they will.

But if evolutionary psychology can provide the *real* reasons for each of our values, tastes, and judgments, where does that leave our sense of self, our conception of what it means to be human? Not in a very good place.

Biology would then become an all-powerful tool, sweeping up the great diversity of human cultures, artistry, beliefs, philosophies, hopes, and fears into a simple biological basket of evolutionary imperatives. Art and music explained not in aesthetic terms, but by their utility in attracting mates. Religion merely an artifact of social bonding in the struggle between competing tribes. Great literature is no longer to be analyzed for plot or style, but explained as pointless narrative that merely stirs the collective unconscious of a bestial past.

What is the good life? What is truth? What is proper, moral, and ethical? In the most extreme of "Darwinian" worlds these are not even questions that matter. Morality itself would be nothing more than a social construction, and our sense of ethical behavior just an evolutionary lubricant greasing the gears of social interactions. What is true and what is right is only that which is of value in the struggle for existence. Developing a philosophy of ethical values would be pointless, because evolution has already placed a powerful set of pseudo-ethics in our heads, a system serving only the ruthless demands of survival and reproductive success.

A DARWINIAN MIND

If the claims of some working in evolutionary psychology were not enough to deflate the human ego, consider the possibility that we are not even in control of our own thoughts and actions. Here the challenge comes from the application of evolutionary materialism to the organ we call the brain. If the brain and the "mind" are one, and modern neuroscience leaves us little choice but to conclude that they are, then our mental selves are creations of the biology of our nervous systems. Those nervous systems, of course, are themselves the products of evolution, shaped by the forces of natural selection.

To author and neuroscientist Sam Harris, this means that free will is an illusion. Our decisions emerge not from conscious choice, but from a series of background forces and mental events over which we have no control. Indeed, even our belief in freedom of action is simply a ruse that evolution has programmed into our brains. As Harris describes it, all we can do is to accept this as fact, "while knowing, of course, that we are ultimately being steered."[33] We are driverless cars running a program we did not write, which we cannot control, and whose existence we are not even wired to sense.

Charles Darwin himself worried about the role of natural selection in shaping the brain. "With me the horrid doubt always arises," he wrote, "whether the convictions of man's mind, which has been developed from the mind of the lower animals, are of any value or at all trustworthy. Would anyone trust in the convictions of a monkey's mind, if there are any convictions in such a mind?"[34] It's a good question. Would we trust a monkey's convictions? Should we trust our own?

If our bodies are merely survival machines, programmed to preserve and propagate the genes within us, then part of that programming, to be sure, is the brain itself. And if the brain is simply a component of that machine, then it serves not truth and beauty, but only a raw calculus of survival and reproductive success.

Let me be clear that I do not believe that the scientific core of evolution negates human belief and conviction as mere byproducts of our struggle to survive. I don't believe that it tells us that our behavior is predetermined or that we lack free will. I don't believe that it reduces us to mere animals, mindless matter, or accidents of nature. Nor does it tell us that our lives are purposeless or pointless.

Our deep, ancestral association with the natural world does not undermine our unique humanness, it's not a knife in the heart of humane intellectual life, and it's certainly not what Robinson once lamented as the "death of Adam." It is, in fact, the best news we have ever received about the world and our place in it. To explain why I believe this, I will first look at the process of evolution itself and grapple with some of the most intimate details of how we came to be. There are many surprises waiting for us there, not the least of which is how evolution allows us to appreciate the actual place we occupy in the scheme of things. We are surely part of Darwin's tangled bank. But we are also the only creatures to be able to transcend it.

Chapter 2

Say It Ain't So

F or many, there is only one way to react to what seem to be the grim pronouncements of the evolutionary priesthood, and that is to reject the whole thing. Indeed, a minor industry now exists in the United States, providing books, videos, lectures, websites, and even glittering museums to support the deniers of evolution. One could be cynical and take the professionals of antievolutionism to be little more than con artists, in it for the money and the bits of prestige that flow their way. But there is much more to it than that. The faithful are not just sheep led astray by slick propaganda and high-tech preachers. Many of their concerns are authentic, their dread of Darwinism runs deep, and their desperation to find something, anything that could disprove evolution is genuine and heartfelt.

Not long ago, I encountered this personal passion firsthand. I had given a lecture about evolution to a crowd of college students and now was taking questions. As a questioner stepped up to the microphone, I could see intensity in his eyes. Most of the questions had been friendly, some asking me to expand on a point, others seeking a comment on something I'd skipped over. But this question was clearly going to be different. Most of the college students in tonight's crowd were supportive, a few were indifferent, and a few were there just to earn a point or two of extra credit from their instructors. When this guy addressed me as "Sir," I knew he was ready with a humdinger.

"If we evolved from monkeys," he said carefully, "why are monkeys still here?" He paused for effect, flashing an icily polite smile.

I smiled back. A few in the audience, mostly faculty, did, too. But after a bit of laughter most of the crowd drew quiet and strained to hear.

I knew the scientific answer, and for a minute, I thought about giving it to him. Evolution doesn't say that we evolved *from monkeys* or from any other organism alive today. Rather, the evidence indicates that we share a common ancestor with all living organisms, and not just with monkeys. But he seemed so certain that he had put this "evolutionist" over a barrel that I decided to have some fun, using a response to the "monkey" question that I and many others have employed over the years.

"I'll answer your question in a moment, but first I have a question for you. Where did Protestants come from?"

"What?" he muttered, almost inaudibly. The smile was still there, but it had weakened a bit.

"No, I'm serious," I insisted. "Where did Protestants come from?" When he hesitated, I decided to offer a little help. "You know, Martin Luther, ninety-five theses nailed to the church door, the Reformation. C'mon, it was in all the papers."

"I guess they came from Catholics," he offered, still puzzled as to why I had asked the question.

"And are Catholics still here?" I responded. Now he saw the point, and so did the audience, as smiles and laughter trickled through the room. "What happened, of course, was that the Christian Church split into two great branches, which today we call Catholic and Protestant," I offered, in a shameless oversimplification of religious history. But no one seemed to mind. The point had been made. "That's also what happened in primate evolution. The branch that led to today's monkeys split from the branch leading to the great apes, of which we are one, many tens of millions of years ago." He sat back down, clearly disappointed with my answer—or, perhaps, with the fact that I had been able to answer at all.

Evolution is not a magic trick in which monkeys change into humans or cats change into dogs. Rather, it's a process in which species diversify and split, continue to diversify, and then split again. The primate lineage to which monkeys and we belong is no exception. It's a mistake to observe that monkeys "are *still* here," since today's monkeys have gone through just as much evolutionary change as we have. So, we do not trace our ancestry back to monkeys or chimpanzees or any other organism living today. Rather, we are part of the great and continuing diversification of life that has touched every corner of this planet.

I'm pretty sure my student friend didn't come to accept evolution just on account of my wiseguy response to his question about monkeys. And I think it's safe to speculate that if he stood up the next year to challenge another speaker or one of his professors on the issue of evolution, he'd have yet another objection. He would, like many, do anything he could to avoid dealing with a concept that would turn his world upside down, the notion that we are descended from mere animals. And I think, to be perfectly frank, he has every right to ask such questions. So, let's see if we can answer them, in advance, right now.

CERTAINTY

How do we know? How can we be sure that the scientific narrative, the story of human evolution, is actually true? Perhaps, as some will say, the Darwinian story is a house of cards. Maybe it's based on nothing more than a couple of bones, a tooth here and there, and lots of wishful thinking. Maybe it's merely a philosophical construct, put together from a desperate desire to elevate "Darwinism" to the status of science. Perhaps, as many hope, evolution just ain't so.

To be sure, nothing in science is ever certain, no theory is beyond challenge, and every bit of scientific evidence should be taken with a healthy dose of skepticism. It is also worth remembering that if we humans are good at anything, it is rationalization. We construct stories that fit our beliefs, and then search mightily for scraps of evidence that might tend to shore them up. Is it possible, therefore, that the evolutionary story is itself based on nothing more than speculation and the hope to find a nonmiraculous answer to the question of human origins? Well, anything is possible. But the story of human evolution is based on firmer stuff than many seem to realize. The logical place to start is right where many seem to think we find only fragmentary evidence of our origins, the fossil record.

A WALK THROUGH THE BONEYARD

No question about it. To most people, the most direct and understandable evidence of human evolution is to be found in the fossil record. What could be more dramatic than the image of an ancient skull held carefully in the hands of an intrepid young scientist? Sure enough, in recent decades, not a year seems to go by without a new fossil discovery breathlessly described as either filling a "missing link" or "revolutionizing" our understanding of human evolution. That sort of publicity-driven hyperbole aside, just how extensive is the prehuman fossil record? Does it truly leave room for doubt?

As a denier, you might start by arguing that the fragmentary nature of that record leaves such a large gap between modern humans and our supposed ancestors that we simply cannot make the clear connection that evolution requires. In other words, why not just deny the existence of any fossil that looks like an ancestor? In a way, this line of criticism harkens back to the day when a literal handful of specimens were all that could be mustered to describe the history of our species. First there was Java Man—but he was just a skullcap. Then there was Neanderthal. Oops! Not an ancestor at all, but a onetime contemporary. Then there was Piltdown Man—a fraud![1] And, of course, there was Peking Man—but all of those bones vanished, suspiciously, during World War II. So where's that missing link?

In truth, *today* we actually confront an embarrassment of riches in the human and prehuman fossil record. Once the first tentative discoveries had been unearthed by intrepid and lucky pioneers such as Raymond Dart and Louis Leakey, it gradually became clear where and how to look for more. As these new specimens moved from the ground into the hands of paleontologists and then into museums and the scientific literature, two things took place. First, the sheer volume of data on human prehistory grew dramatically. Gaps of time and space were filled, links between the present and the past became clearer, and individual specimens were placed into the context of whole populations as the number of such specimens increased. Second, the tools available to analyze that data grew in power and sophistication. Imaging equipment made it possible to analyze complex structures like the skull as never before. Computerized data analysis allowed for more objective comparisons between specimens. And DNA sequencing made possible, in many cases, the sort of direct analysis that allowed for exquisitely detailed comparisons between ancient species and ourselves.

Once, prehuman fossil specimens were so rare that paleontologists felt almost compelled to place them in a straight-line series leading step-by-step to *Homo sapiens*. Before long, however, as the fossil records of one prehuman primate after another became flush with specimens, things changed. Those straight lines turned into branching bushes, and the historical "success" of our species became the tale of a lone survivor from a once diverse lineage.

But could it be that we're not part of a "lineage" at all? Could we simply dismiss the very notion that there is a prehuman fossil connection as nothing more than speculation? Well, fossils themselves are facts, not theories and not speculation. Each individual fossil provides physical evidence of a creature that once existed, just as you and I do today. What are we to make of such creatures if we deny that they are human ancestors or their relatives? At a minimum, we'd have to say that each of these creatures came into existence by some mysterious process at a certain time in the past and then passed into extinction without leaving descendants. More significantly, we would also have to take each and every fossil specimen and sort them, unambiguously, into just two categories: ape and human, with nothing in between. Can we do that? Is there a break in the record large enough to dismiss the entire lineage that way? Let's take a look.

SKULL AND BONE

While the oldest prehuman fossils place the origins of our species in Africa, for many decades scientists have wondered how and when our ancestors moved from that continent into Europe and Asia. A logical place to look, of course, would be somewhere near the intersections of those three continents. As luck would have it, a tiny spot in the republic of Georgia was just such a place.

For decades, archeologists have known the small town of Dmanisi, Georgia, as the site of ruins dating from early medieval times. Excavations there had revealed a number of interesting structures and artifacts, and even an Orthodox cathedral constructed in the sixth century. In 1983, the work of uncovering Dmanisi's history was going along smoothly when a most unusual fossil was unearthed. It was a jawbone fragment from a species of rhinoceros known to have lived more than a million years ago. An ancient extinct rhino in Georgia, nearly 200 kilometers east of the Black Sea? Perhaps the Dmanisi site was even older than its medieval structures indicated.

Indeed it was. As they dug deeper, scientists found more fossilized bones, to be sure, but they found a number of even more tantalizing artifacts—stone tools. Could it be that early humans had once occupied this site? And if they had, just how long ago? The prevailing view had been that humans left Africa no earlier than one million years ago, so it was critical to determine the exact ages of these fossils and tools.

Here the geology of Dmanisi did investigators a welcome favor. There are four layers of solidified volcanic lava at the site, a type of rock that is ideal for age determination. As lava cools, it locks within it both the radioactive isotopes of elements present in the lava and a measurable record of the Earth's magnetic field at the time of crystallization. Both of these can be used to determine the age at which the rock layer was formed. In the case of the Dmanisi fossils, the result was definitive. The fossils and stone artifacts dated to 1.8 million years, comparable to similar specimens found in Africa. So just who had been making these tools?

The answer to that question had to wait until the last day of the excavation season in 1991, when workers at the site spotted a nearly complete jawbone, one with a shape and with teeth that were clearly in a category between modern apes and humans. When the jawbone was first presented at a scientific meeting, the immediate reaction was doubt—doubt and skepticism. Could such a humanlike fossil really date to 1.8 million years, and was it possible that human ancestors had actually moved out of Africa at such an early date? The answer to both questions was yes. In 1999, the first skulls were found at the site, and these were clearly of the type anthropologists called *hominin*, a category that includes both humans and human ancestors. With a cranial capacity of 600 cubic centimeters (cc), smaller than a human but larger than an australopithecine, these fit clearly into the hominin definition. Even more remarkably, it turned out that these skulls were not alone.

Over the next few years, a total of five well-preserved skulls were taken from the Dmanisi site, representing a remarkable collection of individuals living at one place at or around the same period.[2] Other bones were found as well, and it was possible from these to determine the size and stature of these creatures. They stood 140 to 150 centimeters in height,[3] weighed 40 to 50 kilograms,[4] made primitive stone tools, and had brains ranging in size from 550 to 750 cc.

One of these, found in 2005, was particularly striking in terms of its structure and the quality of its preservation. David Lordkipanidze, one of the lead scientists in the Dmanisi group, spent eight years studying "Skull 5," as it came to be called, and reported on it in detail in a 2013 paper in *Science* magazine.[5] In many respects, one of the most interesting aspects of these discoveries has turned out to be the range of morphological variation in the five skulls, even though they were living in the same place at nearly the same time. Had they been found separately, the Dmanisi investigators argued, each might well have been classified as a separate species of the genus *Homo*. Taken together, however, skulls could suggest a view of human evolution in which these and other early *Homo* fossils were part of a single, variable lineage connecting past to present in the history of the human species. Other scientists are not so sure and point out that the process of becoming human involved much more than changes in skull shape and cranial capacity. In fact, an entire "suite" of features is associated with our species, including the distinctive size and shape of the human body, our elongated legs (when compared to other great apes), increased use of meat in our diets, smaller digestive systems, and greater similarity in size between males and females. These features, it seems clear, did not arrive as a single package, but developed separately, leading other scientists to argue for as many as three separate lineages of the genus *Homo* between 1.5 and 2 million years ago.[6] Clearly, the complete picture is not yet resolved. That will have to wait for further discoveries, as questions in paleoanthropology often must.

As fascinating as the stories of these specimens might be, remember the question of denial. Can we use the detailed studies of the Dmanisi skulls to argue that they do *not* support the case for human evolution? Well, the antievolution lobby has already tried to do just that. The Institute for Creation Research (ICR), one of America's largest and most influential antievolution groups, posted an article less than a month after Lordkipanidze's *Science* paper, entitled "Human-like Fossil Menagerie Stuns Scientists." The authors of this article, Brian Thomas and Frank Sherwin, described the Dmanisi finds as nothing more than "human skulls," and proclaimed that as a result species like Neanderthal and Cro-Magnon would "now have to be wiped from the textbooks."[7] They further wrote that *Australopithecus* had now been shown to be "just an extinct ape, and had clearly never evolved into humans." And "without these key players, the popular pageant of human evolution truly should be wiped from the textbooks." So the Dmanisi specimens, including Skull 5, were fully human, according to the ICR, and, therefore, no big deal.

While they were presumably getting ready to edit those textbooks, however, one of the authors may have looked a little more carefully at the actual research report. Within a week he had reversed himself completely, describing Skull 5 as "clearly ape-like" and stating that any interpretation to the contrary "borders on fraud."[8] Suddenly this skull, which previously had been dismissed as just another human, was now being dismissed as just another ape. The very discoveries that had been said to upset the evolutionary story because they were too humanlike were now being dismissed because they were too apelike—and by one of the same authors!

In a previous book,[9] I pointed out a similar problem using fossils discovered well before Dmanisi. Using data put together by James Foley, I presented a list of six hominin fossils and constructed a chart showing how various creationist writers had classified each of them. Not surprisingly, those evolution deniers agreed that none of the fossils was intermediate between apes and humans. The amusing part was that five of the six skulls were described as "apes" by some of the creationist authors and as "humans" by others. As I wrote at the time:

> Which group of creationists is right? I really don't know, and that's the point. In fact, I'm tempted to say they both are. What better proof could one offer of the transitional nature of the human fossil record than the profound lack of agreement of antievolutionists as to how to classify these fossils? Ironically, validation of our common ancestry with other primates comes directly from those who are most critical of the idea.[10]

As the ICR response to the Dmanisi skulls shows, this hasn't changed. For obvious reasons, critics of evolution are determined to find a bright line separating "human" from "ape," and even though they disagree about where to draw that line, they are positive that no creature has ever crossed it.

DRAWING THE LINE

Despite the embarrassing reversal of judgment regarding the Dmanisi fossils, is it possible that the "either ape or human" argument is still valid? Is there data showing a true gap between humans and apes that we could recognize and use to establish the uniqueness of our species once and for all? If there were, surely there should be a way to find that gap quantitatively, a numeric or mathematical technique that would avoid the subjectivity of paleontologists who seem always to be trying to claim that the most critical fossils are the ones they have just discovered.

Actually, there are several ways to do this, and if you and I were truly in the business of evolution denial, surely we would seek to employ such techniques in our quest. An obvious and direct way to do this might be to take published data on every supposedly prehuman fossil and arrange them in a chart based on cranial capacity and fossil age. Surely that would show a distinctive gap if one existed. Several years ago, Nicholas Matzke, then a graduate researcher at the University of California at Berkeley, saw an opportunity to do just that. Noting a paper in the scientific literature[11] listing the cranial capacities of every prehuman skull discovered up to the year 2000, he plugged the ages of each skull and the cranial capacity into a spreadsheet program and graphed the result. Matzke's striking display of this data was remarkably similar to an image published a few years later in a paper analyzing the evolution of the hominin brain.

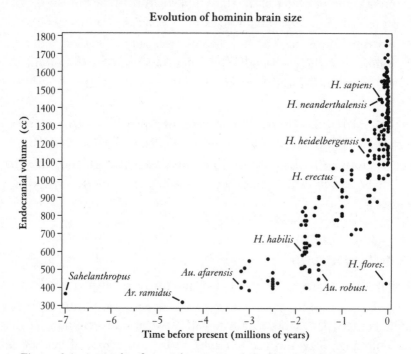

Figure 2-1: A graph of cranial capacity plotted against fossil age for human and prehuman skulls. Each point on the graph represents a specific specimen, and the labels indicate generally accepted scientific names for individual species and subspecies. *Buckner and Krienen, 2013*

Clearly, the difficulty of our task of denial has grown immensely. Using a data set of no fewer than 243 skulls, we find a steady, almost exponential growth in the size of the brain. Beginning with the small 400 cc brains characteristic of australopithecines, we see a broad range of diversity at any given time on the past, leading to the range of brain sizes characteristic of modern humans. Even if we were to accept, for a brief moment, the assertion that members of the genus *Australopithecus* were not human, can we find a clear, objective gap between these organisms and members of the genus *Homo*, our own? As the chart shows, it's just not possible.

Determined, however, to make the best possible case for denial, let's suppose that you point out a limitation of the data. These are nothing more than brain sizes, a single parameter which we already know differs considerably, even among modern humans. We need a more sophisticated method of comparison, perhaps one that includes cranial capacity, but also takes into account the shape of the face and jaw, and properly shows how these vary among modern humans as well as in the fossils we might consider to be ancestors.

Fortunately, the Dmanisi investigators have done exactly such a comparison. In their 2013 paper they placed four of these skulls in a two-dimensional plot where cranial size is represented along the y-axis, and the shape of the skull is characterized along the x-axis. Their results are shown in Figure 2-2.

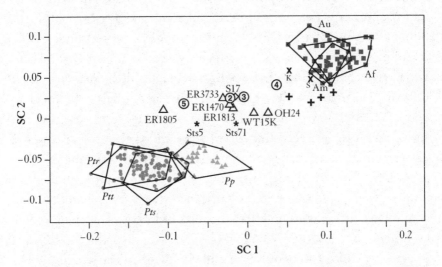

Figure 2-2: A plot of facial shape variation (SC1) versus cranial size (SC2) for the Dmanisi skulls, including data from chimpanzees (*Ptr, Ptt, Pts*) and bonobos (*Pp*) at the lower left, modern humans (including individuals from native populations in Australia, Africa, and the Americas) at the upper right, as well as other *Homo* and *Australopithecus* specimens. *Redrawn, with permission, from Lordkipanidze, D., et al, 2013*

Interestingly, they also included these measurements for chimpanzees and bonobos, as shown at the lower left side of the chart, as well as for modern humans, as shown at the upper right. Their skulls (including Skull 5) are shown on the same chart, along with specimens previously identified as *Australopithecus* and *Homo erectus*. These measurements, while more limited than Matzke's sample of hundreds, allow for a more sophisticated study of the evidence. And the news, I am sure you will agree, is not good for our task of finding a place to draw that line. Not only is there no obvious gap in the data, but the intermediate nature of australopithecines and early *Homo* specimens in the human lineage is glaringly obvious. In fact, the grouped clusters of these specimens fit together almost like stepping-stones in a pond placed edge on edge, moving ever closer to the characteristics of modern humans.

OUR INNERMOST SELVES

While fossils may tell the physical history of our species, in many ways there's an even better story to be found somewhere else—within the human genome. Charles Darwin had a famously incorrect understanding of biological inheritance, the process we now call genetics. In Darwin's view, characteristics from both parents blended together in their offspring, so that each new generation inherited a bit of the previous one. While the passage of beneficial characteristics from parent to offspring might help to account for what he called the "preservation of favored races" in the struggle for existence, it posed a striking problem as well. If a favored characteristic suddenly appeared in one individual, it might well be blended into insignificance after a few generations and would have little chance to effect the sorts of long-lasting changes that evolution requires.

As any biology student learns, unbeknownst to Darwin and just a few years after publication of *The Origin*, an Austrian cleric had solved this problem. Plant breeding experiments carried out by Gregor Mendel and published in 1866 showed that certain traits were controlled by units we now call genes. These genes may come in several forms. For example, in peas the gene for flower color comes in two forms, white and purple. We call these different versions of a gene *alleles*, but that's a minor point. For evolution, Mendel's important finding was that these genes are passed unchanged from one generation to the next. Genes are not "blended," and therefore truly beneficial ones can persist and come to dominate a population, just as evolution by natural selection would require.

Mendel's work lay in obscurity for several decades, but when researchers confirmed that his results were valid for a wide variety of organisms, it became the basis for what we now call classical, or Mendelian, genetics. The emerging science of genetics was then adopted as a key part of the evolutionary mechanism, helping to produce what became variously known as the "new synthesis" or the "neo-Darwinian synthesis" of evolution.[12] As useful as it was for putting scientific flesh on the bare bones of Darwin's work, the new synthesis could not reach full maturity until the physical and chemical nature of the gene was understood. That part of the story began in the 1940s with the identification of DNA as the genetic material, reached a climax when DNA's double-helical structure was unraveled by Watson, Crick, and Franklin in the 1950s, and culminated in the emergence of molecular biology as a distinct field in the 1960s and 1970s.

With the development of a host of new tools and techniques, it is now possible to examine the genomes of living organisms, and to compare them, often in great detail, to one another. Why does this matter? Because it provides a direct way to test the evolutionary notion of common ancestry. Consider this: If two distinct species share a recent common ancestor, it means that both genomes can be traced to a point in the past where the genomes of a single species gave rise to them. This provides any number of ways to test not only whether a common ancestor existed, but even allows us to determine how many years have passed since that single species split into two.

For decades, biologists have applied the tools of molecular biology, with increasing power and sophistication, to questions of human ancestry. At first, these tools were crude and imprecise. They produced, for example, general comparisons of genetic similarity, producing figures like the rough percentages of DNA sequences that two organisms might share. From such studies you may have read that humans share 98 percent or 99 percent of our DNA with chimpanzees, our closest living relatives. Fair enough, but numbers like that don't tell us much about the specific nature of the similarities and differences between these two species, details we'd really like to know. Today, however, we can get right down to the most basic level. The question, therefore, is what such tools tell us about our evolutionary past. The answer, as you may suspect, is that they tell us quite a lot.

A TELLTALE GENE

We can start with a story about human development, one that moves from observations more than a century old to those made only recently with all the power of molecular biology. It's a story about eggs.

We begin our lives, like all animals, as a single cell. Well, actually two cells, sperm and egg, that fuse to form a single cell known (to biologists, anyway) as a zygote. Human eggs, like those of most mammals, are vanishingly small. Unlike the eggs of birds and reptiles, our eggs don't need to contain huge stores of nutrients, because we get the food we need to develop from the bodies of our mothers. In birds and reptiles, as well as in the few mammals that do lay eggs, the story is quite different. They produce large eggs packed with yolk protein and other foods that support the embryo as it grows and develops. To make use of those nutrients, the embryo produces a layer of tissue, called a yolk sac, that completely surrounds the yolk and gradually absorbs those nutrients into its own bloodstream to support its growth.

The embryos of placental mammals, like us, don't have a large store of yolk, so you might think that we have no need to form a yolk sac. But you'd be wrong. Placental mammals, including humans, do indeed form a yolk sac in much the same way that the embryos of birds and reptiles do. The sac, however, pales in comparison to the large sacs of other vertebrates. In placental mammals, like us, there is no yolk to enclose, and the sac is filled with a solution known as *vitelline fluid*. So, why did each of us, as we developed, go through the apparently pointless exercise of constructing a sac that enclosed nothing more than this fluid, whose nutritive value to the embryo is probably very slight? The answer, from the viewpoint of comparative embryology, is that we did this because we share a common ancestor with reptiles that did indeed produce large, yolky eggs. Therefore, we follow a remarkably similar pattern of development ourselves, even to the point of growing, and later absorbing, a small and nearly empty yolk sac (Figure 2-3).

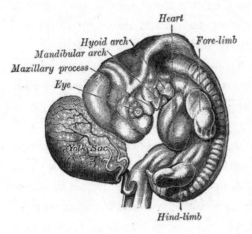

Figure 2-3. A human embryo at 31–33 days of development, as depicted in *Gray's Anatomy*, a classic medical textbook. Note the presence of the yolk sac which, despite its name, does not contain the yolk protein vitellogenin.

Now, it's worth noting that the tissues of the mammalian yolk sac still play a role in development, even if they have nothing to do with that missing yolk. They serve as the source of blood cells in the first month of development and seem to play an early role in transferring nutrients to the embryo.[13] So, the human yolk sac is indeed functional, even if its saclike form is just an evolutionary remnant of the sacs that once enclosed the true yolks of our ancestors. One might argue, therefore, that we should call it something else, since its function has nothing to do with the nonexistent yolk it once enclosed. Indeed, looking at the role the sac plays in mammals today, how can we be certain that its presence in development really does link us to our egg-laying ancestors?

This is where molecular tools help explain something of our true ancestry. One of the essential components of yolk is a protein known as *vitellogenin*. Proteins can help us trace evolutionary pathways because they are coded by genes, which are themselves composed of DNA. So proteins provide us with a pathway inside, into the genome, through which we can look directly at the genetic inheritance of an organism.

Without getting too technical, proteins are produced by stringing together long chains of compounds known as amino acids. There are twenty common amino acids, and they can be assembled in just about any order to produce an almost limitless variety of proteins, from the myosin protein in muscle to hemoglobin in red blood cells to the keratin proteins that make up our hair, fingernails, and skin. One of the primary roles of DNA is to specify the order of these amino acids in proteins. The two strands of a DNA molecule are themselves long chains of compounds known as nitrogenous bases. There are four such bases in DNA, abbreviated A, C, G, and T, and they too can be strung together in any order.

DNA doesn't do this work directly. Rather, when a gene is expressed, its base sequence is copied, or transcribed, into a similar molecule called *RNA (ribonucleic acid)*. Enzymes open up the DNA double helix and produce a single-stranded RNA molecule whose base sequence matches that of the gene itself. That RNA molecule is then used to direct the assembly of proteins, one amino acid at a time, with the aid of a tiny but very complicated molecular machine called the *ribosome*.[14]

Today, we know that some genes code for RNAs that are used for purposes other than protein building, some regions of DNA serve only as regulatory sites, and others are tagged by chemical groups that affect their activity or those of their neighbors. So, with that in mind, let's look at a gene that is essential for egg yolk formation—the gene for vitellogenin.

Since vitellogenin is a protein, there is indeed a gene for it, one that specifies its amino acid sequence. Naturally, the gene is activated in the tissues that produce the prodigious amounts of protein required for the yolk of a large egg. The VIT gene, as it is known, is found in all egg-laying vertebrates, often in multiple copies. But what about mammals, like us, that don't lay eggs? If we are truly descended from animals that once produced large eggs packed with yolk, could remnants of those VIT genes be lurking somewhere in the human genome?

In 2008, a group of Swiss researchers decided to check. Reasoning that what remained of any human VIT gene might be badly damaged by mutations after millions of years of inactivity, they decided to look for it in a clever way. They took note of the genes found on either side of the working VIT genes in the chicken genome and then located the very same genes in the human genome database. Sure enough, those neighboring genes were in the same order in both genomes, and in between them were DNA sequences that could be recognized as the remnants of VIT genes. To be sure, these broken remnants could no longer produce vitellogenin. Some bases in the DNA sequences were missing, and others were changed in a way that had permanently inactivated the gene. What does remain in our genome is a series of VIT *pseudogenes*, relatives of those once-active genes that no longer code for vitellogenin.[15]

Why does the human genome contain these broken traces of genes coding for a protein that we do not make? In evolutionary terms, those pseudogenes are, in the words of the late Stephen Jay Gould, the "senseless signs of history." As Gould writes in one of his marvelous essays, their existence is a direct prediction of evolutionary theory:

> But, Darwin reasoned, if organisms have a history, then ancestral stages should leave remnants behind. Remnants of the past that don't make sense in present terms—the useless, the odd, the peculiar, the incongruous—are the signs of history. They supply proof that the world was not made in its present form.[16]

As the human VIT pseudogenes demonstrate, we were not made in our present form, either. *Proof* is a strong word, especially when applied to evolution, and it is one that Gould used only with great caution. But the striking presence of fossil genes for yolk protein in our genome cannot be explained any other way.[17] The VIT story is particularly compelling because it shows how molecular techniques can confirm an inference made more than a century ago on the basis of morphology alone. But it is not an isolated example.

SCATTERSHOT

Those egg yolk pseudogenes reveal key aspects of our mammalian past, but they don't tell us much about who our closest relatives among those mammals might be. Fortunately, markers abound in our genome that show exactly that with clarity and force.

In Celtic mythology, the land of Tír nan Óg is described as a place of eternal youth, vitality, and joy. A Celt himself, Ian Chambers, of the University of Edinburgh, drew upon that mythology in 2003 when he and his associates chose a name for a key protein they had discovered in embryonic stem cells. For decades, these cells have been intriguing because of their potential to develop into just about any cell or tissue in the body. They are, in effect, forever young, always bursting with the potential of youth. Having discovered a protein essential to that cellular vitality, Chambers gave it the name *NANOG*. The journal publishing Chambers's paper aptly described NANOG as a "new recruit to the embryonic stem cell orchestra," and research on the role of this protein in establishing the identities of embryonic stem cells continues apace.

In the years since its discovery, NANOG has turned out to be just as important as Chambers's group thought, but our special interest in it here revolves around the way in which this gene helps to trace the story of human evolution. In passing along this story, I am especially indebted to Dr. Daniel J. Fairbanks, of Utah Valley University. Dr. Fairbanks, a master painter and sculptor in addition to his scientific work, has written extensively on the history and philosophy of science, and has described much of the molecular evidence for human evolution in two highly readable popular books, *Relics of Eden*[18] and *Evolving: The Human Effect and Why It Matters.*[19]

The functional human NANOG gene is found near one end of chromosome 12. However, as with many other genes, there are extra copies of it scattered around the genome and located on other chromosomes. How do these extra copies get made? One way is from an error in DNA replication in which the same region of a chromosome is mistakenly copied twice. This clearly happened at least once to part of chromosome 12, since there is an imperfect copy of NANOG located right next to the working copy. However, there's another way these imperfect copies, or pseudogenes, can be produced. As we've seen, when a gene is expressed, its DNA base sequence is first copied into a complementary sequence of bases in an RNA molecule, which is then used to direct the production of a protein. So far, so good. But sometimes, very rarely, enzymes within the cell copy those RNA base sequences right back into DNA, producing yet another copy of a gene like NANOG. These copies of RNA, which are known as *processed pseudogenes*,[20] can be inserted just about anywhere in the genome, on any chromosome. There are nine of these processed NANOG pseudogenes scattered on seven different human chromosomes, with locations every bit as random as if they came from the pellets of a shotgun blast.

Here's where it gets interesting. What would you say if you came across two different species, each of which had these very same nine scattershot processed NANOG pseudogenes in matching places on the very same chromosomes? Coincidence? Well, maybe once or twice, but nine times in positions that match up perfectly on the chromosomes of two different species? That simply wouldn't be possible—unless they both inherited that pattern of pseudogene location from a common ancestor. Then it would make perfect sense. If those pseudogenes had been inserted, one or two at a time, over millions of years, first into one chromosome and then into another and later into yet another, a pattern of such genes on several chromosomes would have developed in that species that was every bit as specific as a fingerprint. Then, when that species split into two, each of those daughter species would have retained that fingerprint by inheriting the identical pattern of NANOG pseudogene locations among their chromosomes. That pattern would persist for millions of years, even as these two species went along their separate evolutionary pathways, specializing in ways that made them increasingly different and distinct from each other. Mutations would accumulate, a new pseudogene or two might be inserted into the genome of one of the species, but the matching pattern would remain, revealing that many millions of years ago the two species had indeed shared a common ancestor.

Those two species are humans and chimpanzees (Figure 2-4). But the story is even better than you might suspect from the figure alone.

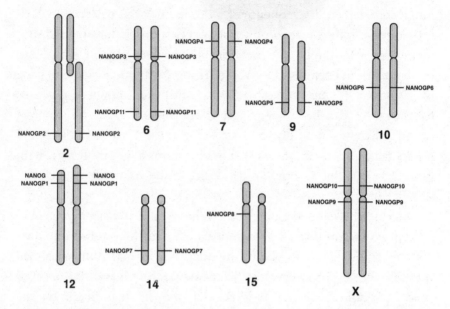

Figure 2-4: The approximate location of the "Daughters of NANOG," the processed pseudogenes scattered randomly throughout the human and chimpanzee genomes. Note that the locations of these randomly inserted sequences match on all but chromosome 15, which contains a NANOG sequence found only on the human chromosome. (Human chromosomes are shown at left, chimpanzee ones on the right.) *Redrawn, with permission, from Fairbanks, 2007*

If you look closely at the map of human and chimpanzee chromosomes, you will see there's one pseudogene that doesn't match up. It's located on human chromosome 15, and there is no corresponding copy in the chimpanzee genome. So, where did it come from?

By counting the number of mutations that a pseudogene has accumulated compared to an original working gene, it's possible to estimate, with some accuracy, when the pseudogene was formed. Lots of mutations? Then the pseudogene is pretty old. If there are only a few, then it was formed more recently. NANOGP8, as this pseudogene is known, is very young, having been formed only about 2 million years ago.[21] By contrast, the other nine processed NANOG pseudogenes all reveal ages greater than 22 million years. NANOGP8 is missing from the chimpanzee genome because it arose only in the human line, at a time when the ancestors of our species had already split off from the ancestors of today's chimpanzees. As one would expect, the only NANOG pseudogene that differs between humans and chimpanzees is the newest one, a pseudogene produced after the two species had split into separate lines from their common ancestor, roughly 5 million or 6 million years ago.

NANOG, the gene that draws its name from the Celtic land of eternal youth and vitality, is essential to the renewal of cellular vigor that accompanies each new human generation. But the "eleven daughters of NANOG," as two researchers[22] once called these pseudogenes, stand as markers to a past many generations distant from our own, a past we share with our closest relatives on planet Earth. Neither we nor they, as different as we have both become, were made in our present forms. But written deep within us is a history and kinship we cannot deny.

FUSION

NANOG is just one example of a pseudogene family documenting the common ancestry of humans and other primates. There are many others. But there's an even simpler way to make the very same point. So simple, in fact, that you can tell the story with just one picture, and here it is:

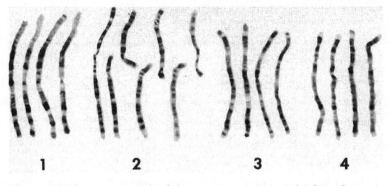

Figure 2-5: Photomicrograph of chromosomes 1, 2, 3, and 4 from four primate species, humans, chimpanzees, gorillas, and orangutans, arranged left to right in order. *Yunis & Prakash, 1982*

In 1982, two scientists took microscopic images of human chromosomes and compared them to the chromosomes of three other great apes—gorillas, chimpanzees, and orangutans. They cut out the corresponding chromosomes from photographic prints and lay them side-by-side for comparison. Not surprisingly, they were nearly identical, even when they were stained to make their internal banding patterns visible. To be sure, there were a few differences. On a couple of chromosomes, there were inversions, places where part of a chromosome had been flipped around to wind up in reverse orientation. But among these slight differences there was one really striking one. Human chromosome number 2 seemed to have no match among the chromosomes of these three other primates. How come?

I'm pretty sure the authors of this study, Jorge Yunis and Om Prakash, of the University of Minnesota, weren't baffled by this apparent discrepancy for more than a New York minute. That's because there were two chromosomes in the great apes that didn't seem to have a match in the human genome, either. Since chromosomes are numbered according to their size, these would have been called chromosomes number 12 and 13 in their respective genomes. But Yunis and Prakash realized immediately they were a perfect match for the top and bottom halves of human chromosome 2. In a flash, they had solved a tiny mystery of human evolution. Our species, you see, has 46 chromosomes arranged in 23 pairs. Each of us inherits 23 chromosomes from our mother and 23 from our father. However, all the other great apes have 48 chromosomes, or 24 pairs. So, if we do indeed share common ancestry with these other primates, why are we missing a chromosome?

Well, as these scientists realized, we aren't missing a chromosome at all. Rather, in the course of human evolution, two chromosomes that are still separate in our primate relatives had fused together to become a single chromosome, so that modern-day humans have 23 pairs instead of 24. No loss of genetic information, just a rearrangement of the sort that happens from time to time in many organisms. The way in which this fused chromosome matched up with two separate chromosomes in the three other species enabled Yunis and Prakash to call their study a "pictorial legacy" of the "origin of man." And so it was. But there's a little more to the story.

Did a fusion between two separate chromosomes really happen the way these scientists assumed? In the decades since their study, the technical tools emerged to test their ideas. We can now look directly at the DNA base sequences of the supposedly fused chromosome and test the idea of common ancestry.

If human chromosome 2 really did come about this way, it should have telltale signs of its origin as two separate chromosomes in the common ancestor of the great apes. Every chromosome has a *centromere*, a more or less central region where a slight constriction is sometimes visible.[23] If our second chromosome resulted from a telomeric fusion, then it should have two centromeres, or at least the remnants of two centromeres, one each from the ancestral chromosomes. While Yunis and Prakash couldn't tell this from their microscope images, in 2005 a detailed DNA sequence study[24] of chromosome 2 found both the predicted centromeres. In fact, it did even more than that, identifying the DNA sequences surrounding both centromeres as corresponding to the centromeres of what had been previously identified as chromosomes 12 and 13 in the chimpanzee. This not only confirmed the identities of the two great ape chromosomes corresponding to the two ends of human chromosome 2, but also supported the fusion hypothesis.

Even stronger support came from another line of molecular evidence. The ends of human chromosomes, the telomeres, have a particular DNA base sequence, TAAGGG, repeated hundreds of times. Now, if human chromosome 2 had indeed been formed by a fusion of two chromosomes, two groups of telomere sequences should have been squashed together right at the point of fusion, one from each of the original chromosomes. In 1982 (the time of the Yunis and Prakash study), there was no way to check this. But once the human genome, including chromosome 2, had been completely sequenced via the Human Genome Project, it was possible to take a look. Today we can read the DNA base sequence of nearly the entire chromosome and look for those telomere repeat sequences where they wouldn't normally be found, right in the middle of the chromosome. Sure enough, there is a region in the middle of the chromosome with roughly 150 of these telomere repeats, right at the fusion point. In fact, it's even possible to pinpoint the exact base where the fusion took place: 113,602,928 bases from the end of the chromosome. (Note: I've included a more detailed treatment of the nature of the fusion site in a technical appendix at the back of the book.)

When I speak to lay audiences on evolution, I almost always bring up the story of our second chromosome. There's nothing quite like it in terms of its simplicity and explanatory power. First, I point out that odd discrepancy of chromosome number between our species and the great apes: 46 for us, 48 for them. Next, I ask what the claim of common ancestry would require to explain this mismatch. Quickly, they realize that only one explanation is consistent with evolutionary common ancestry. That is that one of our chromosomes must have been formed by the fusion of two chromosomes still separate in the other primate species. With that prediction in hand, we can then ask, "What would such a chromosome look like? And how would we recognize it?"

As we've seen, such a chromosome should have head-to-head telomere repeats at the fusion site and gene sequences on either side of the fusion site that match those of the two chromosomes in primate relatives, like the chimpanzee. My next question is, "Do we have such a chromosome?" My answer: "We sure do. It's human chromosome number 2, which has every one of the markers I've just described." Case closed.

I'm always struck by the effect the chromosome 2 story has on evolution skeptics. Suddenly, they seem to realize that common descent is not a fanciful guess fashioned from the desires of secularists to undermine their faith or destroy Western civilization. Rather, it is a straightforward conclusion drawn from the hard evidence of DNA sequences in the human genome. Let's just say it makes quite an impression.

DENIAL'S BURDEN

So, now we return to that opening question. Can we say it just ain't so? Can we take the obvious physical similarities between humans and the other great apes and say they just don't matter? Can we dismiss the growing and increasingly detailed fossil evidence as illusion or deception? Can we find refuge in the overwhelming complexity of the genome, even while ignoring the marks of common descent that are found everywhere throughout that genome? I think the answer is clear.

Of course, we can't. The intellectual burden of denying human evolution in the face of so many lines of evidence would be far too great for any fair-minded person to sustain.

To be sure, we don't know every stop along the evolutionary pathway, and we will surely be surprised by much of what we learn in the years ahead. But the broad outlines of our story are clearer than ever before. We are creatures of evolution, with all that implies about our bodies, our minds, our thoughts, our hopes and dreams. And now we can begin to ask the really interesting question. What does it mean to be a creature fashioned by the hand of evolution?

Chapter 3

Chance and Wonder

Evolution means that we find our origins in the natural world. It tells us that we were not here to witness the formation of the Earth, the origin of life, the rise of plants and animals, or even the end of the dinosaurs. It means that we, as human creatures, are newcomers to this planet of life. And, perhaps most significant, that we can trace our ancestry back to a point at which it merges with every other living thing. Is that really so awful to contemplate?

Might it be possible instead to look at the whole process, especially our recent arrival, as the culmination of life's plan? If we are the crowning glory of life on Earth, perhaps all that came before was destined to gradually construct the present living world. What would be wrong with that? Why can't we look at evolution that way and thus recover the sense of destiny and entitlement with which we once viewed ourselves? To understand why so many would find this troubling, we have to see how the narrative of human origins came to be and why it has changed over the years.

FIRST GLANCES

The idea of human evolution did not emerge all at once. In fact, it was centuries in the making. It began, as many scientific breakthroughs do, in the innocence of curiosity. Isaac de la Peyrère was a Frenchman fascinated by the discovery of certain stones, apparently sharpened by chipping and scraping. He recognized them as tools, possibly arrowheads, and attributed their production to primitive people living in an era *before* Adam and Eve. If this were true, he realized, it might explain away several nagging problems that emerged from a close reading of the Genesis narrative. Where, as Sunday school children have always asked, did Cain, the firstborn son of Adam, find his wife? Whom did he fear after being exiled for killing his brother, Abel? And even more perplexing, who would have been around to take residence in the "city" that Cain built and named after his own son, Enoch?[1] As de la Peyrère pointed out, the existence of other peoples, unrelated to Adam and Eve, would solve these problems at a stroke.

But the reaction of his countrymen to de la Peyrère's helpful suggestions about Genesis was not good. His book, *Men Before Adam*,[2] was publicly burned in 1656, just a year after its publication. Peyrère, fortunately, was not, although he was coerced into renouncing his ideas before the Pope in Rome. His ideas, however, persisted and were taken up by others equally curious.

In 1771, Johann Friedrich Esper discovered human remains associated with the bones of extinct cave bears near Bamberg, Germany, leading him to doubt the Genesis story. In the early 1800s, John MacEnery, a Catholic priest, discovered human bones associated with those of extinct animals in ancient sedimentary rocks along the coast of England. As Father MacEnery came to realize, humankind seemed to be far older than the Genesis narrative suggested.

In the scientific sense, an even more troubling line of inquiry had been set in plain view even earlier. This was a book, *Systema Naturae*, first published in 1735 by the Swedish naturalist Carolus Linnaeus.[3] In it, Linnaeus set out the first truly systematic classification of living organisms. It is from him that we derive the system of two-part scientific names used even to this day. Following Linnaeus's system, green beans are known everywhere as *Phaseolus vulgaris*, horses are *Equus caballus*, and we humans are *Homo sapiens*. Convinced that his work revealed a Divine plan in nature, in the frontispiece of his book he wrote, *"Deus creavit, Linnaeus disposuit."* Loosely translated, "What

God has created, Linnaeus has classified." In his own mind, here was a man doing nothing more than describing the glories of Creation. But in so doing, he had also laid the groundwork for dramatic downgrading of the species so kindly described by Scripture as just "a little lower than the angels."

The crowning glory of that creation, of course, was man himself, and to reflect man's special place, Linnaeus made it clear that we belonged to the "first" or "highest" group of animals. He took the Latin word *primas* (*primatis*), meaning "noble" or "of the first rank," and coined a term we still use today. Humans belong to the "primates," the "first animals." Today we might find that term ironic, since in the Middle Ages the term *primate* was often applied to Church officials of high rank. But I doubt very much that Linnaeus saw any irony in his use of the term. He simply took the view that we stood at the very top of the scale of living things, and that's how he classified us.

Linnaeus's scheme confirms our place at the pinnacle of the living world, except for one detail. We were not alone among the animals he called primates. Here Linnaeus's stubborn reliance on scientific principle betrayed him. We humans, *Homo sapiens*, were lumped in with apes and monkeys in a subcategory of primates called *Anthropomorpha*, which means "manlike." Right-thinking naturalists of the time objected. How could he possibly compare the sublime soul of noble humanity to such beasts? Responding to one such critic, Johann Gmelin, Linnaeus wrote:

> You are not pleased that I've placed Man among the *Anthropomorpha*, perhaps because of the term "with human form," but a man knows himself. Let's not quibble over words. The name we use does not matter to me. But what I ask of you and of anyone else in the whole world is to cite a single generic difference between man and ape based on the principles of Natural History. I certainly know of none. If only someone might tell me a single one![4]

Then, knowing who might have been looking over his shoulder, Linnaeus confided certain fears to Gmelin:

But, if I had called man an ape, or vice versa, I should have fallen
under the ban of all the theologians. It may be that as a naturalist
I ought to have done so.[5]

It may have been, indeed. But Linnaeus knew who held power in the
society of his time. He persisted with his classification, and in later edi-
tions even removed sentences implying the fixity of species. Nonetheless, he shied away from speculation as to human origins, as did many
other naturalists of the time. The tenth edition of his great work appeared in two volumes, published in 1758 and 1759. The frontispiece of
that edition, which guided biologists throughout much of the great age
of exploration, contained lines surely guaranteed to please those "theologians."

O Jehova.
How great are Thy works!
How wise Thy deeds!
The Earth is filled with your wonders![6]

Nonetheless, Linnaeus's insistence on the biological kinship of humans
and other primates made it almost respectable for European scientists
to consider this relationship, and consider it they did. Almost exactly
one hundred years after the publication of that tenth edition, Charles
Darwin and Alfred Russel Wallace would take the observations of Linnaeus a critical step further. Species were not fixed. New species origi-
nate by descent with modification from older ones. And the process of
natural selection continually shapes the characteristics of species. While
neither Darwin nor Wallace initially applied these principles to human
origins, neither saw fit to exclude the human species from the process of
evolution, either. In *The Origin of Species*, Darwin coyly noted that as a
result of his theory, "Light will be thrown on the origin of man and his
history." For the moment, the author of *The Origin* left it to his readers
to decide what that light might reveal.

But Thomas Henry Huxley, Darwin's great popular champion, did
not hesitate. In a series of lectures and papers, Huxley laid out his own
interpretations of human ancestry, summarizing them in an 1863 book,

Man's Place in Nature. In it, Huxley found " . . . no rational ground for doubting that man might have originated, in the one case, by the gradual modification of a man-like ape; or, in the other case, as a ramification of the same primitive stock as those apes."[7]

As Huxley lectured, delighting in the public spotlight, Darwin worked quietly, and before long he was ready to break his silence with a full-length tome on human evolution. That book, *The Descent of Man, and Selection in Relation to Sex*, was published early in 1871, and its first printing sold out within three weeks. Unlike Huxley's brief and narrowly targeted arguments for human descent, Darwin's work was expansive. It dealt with sensory organs, hair, muscles, and reproductive organs. It analyzed human variability, considered the courtship rituals of butterflies, speculated on the coloration of fishes, and described sexual dimorphism in spiders. The essence of this long argument was that humans were, in every respect, animals belonging to the same group as other primates. As such, we were and had been subject to the same forces of variation and natural selection as other animals, so there was every reason to believe that our species, like all others, was the product of evolution. We weren't just similar to other primates, as Linnaeus had described, we were their *cousins*, their blood relatives, their kin. In a Victorian age where hereditary lineage mattered deeply, the impact on popular culture was immediate.

A few clever pundits, taking stock of human failings, pretended that our primate relatives should be the ones to take greatest offense at this outrageous suggestion. Along these lines, in 1871 the great political cartoonist Thomas Nast took a break from depicting the scandals of New York City's Tammany Hall to take a tongue-in-cheek shot at the author of *The Descent of Man*. In a *Harper's Weekly* cartoon, Nast drew a tearful, disconsolate gorilla pointing an accusing figure at Darwin while a bystander asked, "Now, Mr. Darwin, how could you insult him so?"[8]

Other cartoonists eschewed irony and ridiculed the very idea that humans might be descended from other animals. *Punch*, the British humor magazine, featured a father reading from *The Descent* to his horrified wife and daughter.[9] The *Hornet* drew Darwin himself as a hairy ape, lumbering on all fours,[10] while the French magazine *La Petite Lune* went even further, showing a bearded, apelike Darwin swinging from a tree branch.[11]

Darwin's offense was obvious. He had given our species's self-image its biggest shock ever. The Adam of Scripture might not yet be dead, but he was surely on life support.

Darwin's argument in *The Descent* was based on comparative biology. Powerful as his argument was, however, there remained a striking problem with it: the absence of so much as a single fossil to document human origins. In the beginning of the book's sixth chapter, Darwin admitted this difficulty:

> " . . . the facts given in the previous chapters declare, as it appears to me, in the plainest manner, that man is descended from some lower form, notwithstanding that connecting-links have not hitherto been discovered."[12]

While Darwin seemed to say in this passage that fossils were not necessary to make the case for human evolution, a bit later in the same chapter he returned to this question and did a bit of serious hand waving:

> With respect to the absence of fossil remains, serving to connect man with his ape-like progenitors, no one will lay much stress on this fact, who will read Sir C. Lyell's discussion, in which he shows that in all the vertebrate classes the discovery of fossil remains has been an extremely slow and fortuitous process. Nor should it be forgotten that those regions which are the most likely to afford remains connecting man with some extinct ape-like creature, have not as yet been searched by geologists.[13]

In a sense, he told his readers, no matter that we don't yet have prehuman fossils. Heck, we don't have them for most other vertebrate classes either, so what's the big deal? At the time, however, I suspect Darwin knew it was a big deal, and that's why he took care to address the issue. Facts matter, and despite the care with which he had assembled evidence attesting to our close relationship with other primates, one critical piece of evidence was still missing—a genuine fossil that might link us to one of our prehuman ancestors. As if to admit that to his readers, Darwin was bold enough to suggest exactly where to look:

> In each great region of the world the living mammals are closely related to the extinct species of the same region. It is therefore probable that Africa was formerly inhabited by extinct apes closely allied to the gorilla and chimpanzee; and as these two species are now man's nearest allies, it is somewhat more probable that our early progenitors lived on the African continent than elsewhere.[14]

Had he been wrong, it is difficult to say what would have happened to Darwin's theory. But he was not.

STORIES IN STONE

Well before he finished work on *The Descent*, Darwin and his associates were aware of the discoveries of humanlike fossil bones in several places throughout Europe. The most famous of these were first found in the Neander Valley of Germany. These Neanderthal fossils (*Homo neanderthalensis*) awakened the imagination of naturalists to the idea that evidence for a "missing link" between apes and human might well be found right beneath their feet. Alas, it soon became clear, as Huxley himself pointed out, that this was not the case for this particular find. In an 1862 paper,[15] Huxley maintained "In no sense . . . can the Neanderthal bones be regarded as the remains of a human being intermediate between Men and Apes." As Huxley recognized, Neanderthals were far too close to modern humans to be regarded as true intermediates, and so the search had to go on.

Darwin himself may well have wondered whether those "ape-like progenitors" would ever be found. In 1868, he drew a sketch of primate evolutionary relationships, depicting his view that humans had split off from the other great apes well before the emergence of modern forms of chimpanzees, gorillas, and orangutans.[16] When Darwin died in 1882, his visionary theory of human ancestry had yet to be confirmed by so much as a single fossil specimen. Within a decade, however, that would change.

If Neanderthal was not quite the specimen Darwin might have hoped for, other candidates soon appeared. The most important of these consisted of a skullcap, a tooth, and a femur found on the island of Java by the Dutch naturalist Eugène Dubois in 1891. Although Dubois's finds were controversial for decades, we now regard them as examples of a species called *Homo erectus*.[17] Subsequent discoveries provided many other examples of the same species at a variety of locations around the world. Significantly, the cranial capacities of these specimens range from roughly 850 cc (cubic centimeters) to 1,100 cc, allowing them to be thought of as intermediate between apelike ancestors (close to 400 cc) and modern humans (ranging from 1,200 cc to 1,500 cc). Although not everyone agreed, for the first time it was possible to argue that a missing link had indeed been found. The dam had been broken.

To be sure, there were some false starts, and one is routinely put forward as evidence of scientific duplicity on matters of human ancestry. In 1912, an amateur archeologist found portions of a skull and a jawbone in a gravel pit near the English town of Piltdown. These bones were eventually turned over to Arthur Keith, a respected anatomist, who pronounced them genuine. The Piltdown discoveries engendered great publicity in Britain, and were even immortalized in a 1915 painting by artist John Cooke. In that painting, Keith dramatically measures the skull of Piltdown man, surrounded by its discoverers, while a portrait of Charles Darwin hangs none too subtly in the background. Piltdown's brain was almost exactly the size of a modern human's, with a jawbone strongly resembling that of an ape. While many in the scientific establishment rejoiced that "the first Englishman" had been found, others were troubled by the obvious mismatch between skull and jawbone. In truth, Piltdown did not make sense as a true intermediate, and the validity of the find was called into question almost immediately by paleontologists like Marcellin Boule in France and Gerrit Smith Miller in the United States. As early as 1915, Boule and Miller both insisted that the jawbone had, in fact, come from an ape, and they turned out to be right. Confusion over the fossils was resolved in 1953 when radiometric dating techniques were employed. Piltdown was indeed a hoax. The skull was that of a modern human, while the jawbone was indeed from an ape, with its teeth filed down to make them appear somewhat more human.

Maybe it was time, at last, to look in Africa.

By the time the Piltdown fraud had been discredited, the first great find in Africa had been made, almost by accident. Raymond Dart, an Australian, had just been hired as head of the anatomy department of a South African university. When one of his students presented him with an unusual primate fossil from a limestone quarry, Dart wondered if the quarry might contain other interesting specimens. He asked for and received two boxes of broken limestone from the quarry and proceeded to comb through them. Out of the second box, Dart removed a series of skull fragments and carefully pieced them together to form the face and skull of a young child. Dart named his specimen *Australopithecus africanus*, the "southern ape" from "Africa." His 1925 paper in the journal *Nature* bore the subheading "The Man-Ape of South Africa."[18] At long last, Darwin's suggestion had borne fruit. Adult specimens of *Australopithecus* were soon found throughout Africa, many of them by Dart's colleague and supporter, Robert Broom.

While *Homo erectus* was nearer to modern humans in terms of brain size, the cranial capacity of *Australopithecus africanus* was only about 500 cc, much closer to that of modern apes. At long last, elements of a whole chain of missing links were beginning to fall into place. At the end of the Second World War, focus on the African continent increased, and well-organized teams of fossil hunters began to search through promising geological regions. Time and time again they struck pay dirt, and discoveries with names like *Homo habilis*, *Paranthropus*, and *Zinjanthropus* popped up on the pages of newsmagazines. Increasingly accurate radiometric analyses of volcanic material associated with these fossils added to the story by placing these specimens in chronological order.

By 1965, Time-Life Books was emboldened to publish a glossy, coffee-table book with the title *Early Man*.[19] At the beginning of one of its chapters anthropologist F. Clark Howell stated flatly, "It is now a proven scientific fact that man was millions of years in the making." Any hope that a gap might remain to separate us from the rest of the animal world was fast disappearing. One by one, the missing links were being found, and we were on our way to becoming just another critter in the jungle of life.

A ROYAL ASCENT?

There was, however, one way in which humans might have taken some comfort from that emerging story, and it was right there in that Time-Life book. In a chapter entitled "The Road to *Homo sapiens*," a dazzling five-page foldout depicted a 25-million-year ascent from proto-apes to true apes to australopithecines to early versions of *Homo* to Neanderthal and finally to modern humans. Each of the fifteen figures in *The March of Progress* was male, each confidently stepped left to right, and each did so with his right leg extended so as to obscure a certain part of the male anatomy. This striking image was the work of Rudolph Franz Zallinger,[20] a celebrated artist who had turned his talents toward illustrations of the evolutionary past. It was a classic figure and would be reproduced in textbooks, parodied countless times, and adapted to sell everything from computers to soft drinks and beer.

In its own way, the figure confirmed a certain pleasant conceit about the human species. In *The March of Progress*, we humans stood at the end of this rising series, the proud culmination of a process that step-by-step had made us taller, stronger, smarter, and even better looking. Maybe we really had evolved from apes, and it was time to swallow the bitter pill of that lowly ancestry. But in so doing, perhaps we could recover more than a little bit of pride by placing ourselves at the very pinnacle of life. It was one thing to admit we had evolved, but would that be so bad if the entire evolutionary process was destined to produce *Homo sapiens* as its crowning work?

Perhaps we could actually think of our emergence as the ultimate achievement of the evolutionary process, preordained from the very beginning of time. Our planet was born in fire, richly endowed with the materials of life and the sparks of energy. These forces kindled the appearance of life, first simple, then complex. That life spread throughout the planet, changing and adapting, exploiting habitats and devising new ways to survive in virtually every environment on Earth. Eventually, inevitably, life produced creatures with greater and greater awareness and mental acuity, until finally *Homo sapiens* emerged to master, harness, and rule the marvelous diversity of living things. We are the omega, the final products of evolution, the ultimate goal of billions of years of evolution. We've arrived.

In 1965, that message seemed to fit the times perfectly. New York was in the midst of a great World's Fair that celebrated progress at every turn. The General Electric Corporation had produced a dazzling pavilion celebrating the ways in which its products were making our lives better and better. Other companies depicted a brilliant future with colonies on the moon, arid deserts transformed into productive farms, and laborsaving devices to make life easier, safer, and more creative. Exhibits from countries around the globe endorsed this great vision, each wrapping it in the cloth of its own culture and national identity.[21]

In the smiling optimism of that World's Fair, nothing seemed beyond the reach of progress. So it was only logical to ascribe our own ascent to a natural progress built into the character and history of the living world. Every day in every way, things were getting better and better. America herself had naturally risen to the top rank among nations as a result of our pursuit of democratic ideas, good government, science, peace, and freedom, and this seemed to be the natural order of things. Human evolution could be understood the same way. We were truly the ultimate stop on the carousel of biological progress.

Before long, however, even this would be taken from us. The World's Fair would end, and its site would start a long, slow decline into disuse and neglect. America's primacy and self-confidence would come up against words like *Vietnam* and *Watergate*. And the hope that we could claim a triumphant spot at the top of the evolutionary tree for ourselves would gradually fade away.

There were hints, even on the glossy pages of *Early Man*, that all was not a steady rise to the top. Above those images of confident primates striding into the glorious present were a series of bars indicating the ages in which each of the several forms had existed. More than a few of them overlapped before extinction claimed them, indicating that several of these species had actually lived side by side. The process of human evolution was not a steady, straight-line transformation. Before long, new discoveries would only add to the complexity of our family tree, and waiting for us amid those complexities would be one of life's little jokes.

THE LAST HUMANS?

If I were asked to recommend an updated companion to *Early Man*, my choice today would be *The Last Human*,[22] another large, attractive book suitable as a conversation piece in any living room or den. The book's subtitle is *A Guide to Twenty-Two Species of Extinct Humans*. Its stated aim is to bring us closer to an understanding of the "colorful precursors and relatives" of *Homo sapiens*, and it does so brilliantly. A team at the American Museum of Natural History in New York began with fossil casts and carefully reconstructed the faces of a series of vanished hominids by layering on muscle, tendons, and skin with strikingly realistic results. Full-color photos of these reconstructions grace the pages of the book, many depicted in their African habitats. Even more compelling, however, is the story it tells with respect to our own ancestry.

Human evolution is anything but a step-by-step rise to modern greatness. In fact, human history is "the story of a diverse hominid family that sprawled rather than snaked across the ages,"[23] in the words of Ian Tattersall, who wrote the volume's introduction. More than four decades of research and discovery separate the publication of *Early Man* from *The Last Human*, and it is now clear that our present age, in which a single hominid species survives (our own), is a radical departure from a not too distant past in which multiple species walked the plains and forests of Africa. Or, as Tattersall put it, "Our species is more like the sole surviving twig on a luxuriantly branching tree that represents a story of constant evolutionary experimentation—and, as often as not, of extinction."[24]

This matters because of our own willingness, maybe even our compulsion, to view human emergence as a story of triumph and "success" in the evolutionary scramble. In one important respect, it is nothing of the sort. Stephen Jay Gould explored this tendency in an essay published in his 1991 collection *Bully for Brontosaurus*.[25] Gould's example was one of the most storied and beloved of all domesticated animals, the horse. Today, looking at the role the horse has played in human history, including the triumphs of horseback-riding Spanish conquistadores over Native Americans, one could view these magnificent animals as one of evolution's great success stories. "A horse, a horse, my kingdom for a horse," in the words of Shakespeare. Indeed, imagine how human history might have changed had the tribes of the Americas been mounted on these beasts when Europeans first arrived.

In that essay, "Life's Little Joke," Gould describes an 1876 encounter between Thomas Henry Huxley and Yale paleontologist O. C. Marsh. Huxley had earlier taken a series of European fossil horses and attempted to arrange them in a continuous series that would document the evolution of the modern species. But even his best attempts failed for reasons that Marsh was about to make clear. Modern horses, as Marsh's extensive fossil collection demonstrated, had actually evolved in North America, not Europe. At several times in ages past, certain species of horses had migrated to the Eurasia landmass, but these specimens provided only a fragmentary and incomplete record of equine evolution. Once shown the evidence, Huxley was quick to agree to Marsh's claims. Marsh then drew for Huxley a ladder of progress showing how small horses with three toes had evolved into much larger contemporary horses with a single toe modified to form the hoof. Huxley used this chart in lectures as he continued his American tour, while later versions included skulls of the fossil horses, correlated with the geological time periods in which each had been discovered.[26]

The impression given by these illustrations was very much like the iconic Time-Life *March of Progress*, of a rise from ancient simplicity and obscurity to modern greatness. But that impression was wrong. As the richness of the equine fossil record was further explored, it became apparent that the true story was not one of a linear pathway, but rather a bush with multiple overlapping branches that split time and time again. These branches were pruned multiple times by extinction, until only one genus was left to survive into the present—the modern horse, *Equus*. More recently, Bruce MacFadden, of the University of Florida, has produced a worthy successor to Marsh's illustrations, a diagram depicting more than thirty extinct genera and citing more than a hundred extinct species.[27]

Why, then, were brilliant scientists such as Huxley and Marsh led to depict equine evolution as they did, showing a single pathway rising triumphantly from fossil ancestors to the modern species? The reason, as Gould pointed out, was not the success of the horse lineage, but rather its near utter failure.

Truly successful evolutionary lines branch repeatedly, filling the world with dozens, even hundreds of species. Think of the most successful mammalian lineage, the rodents. With more than two thousand species, rodents account for nearly 40 percent of all mammals. As a result, no one is tempted to trace a single line of ancestry back into history for the Norway lemming (*Lemmus lemmus*) or the prairie vole (*Microtus ochrogaster*), because any attempt to do so intersects with a tangle of living branches related to scores of other living species. Not so with *Equus*, the sole surviving representative of an evolutionary tree that has all but been extinguished. Today we have only a handful of species in that one remaining genus of what was once a great and dynamic evolutionary family. The great irony of this, of course, is that taken as a whole, the horse lineage was remarkably unsuccessful. All but one of its many branches perished, and even that one represented nothing more than a few struggling species until a bipedal primate decided that some of these animals might be useful for its own purposes.

That, according to Gould, is life's little joke. Unsuccessful lineages are the only ones that tempt us to see evolution as a ladder-like progression from past to present. Over time, evolutionary change actually produces a bush of branching lines as species split and split again without hint of progress or direction. The notion of evolutionary progress, according to Gould, can "only apply to unsuccessful lineages on the very brink of extermination—for we can linearize a bush only if it maintains but one surviving twig that we can falsely place at the summit of a ladder."[28]

Today's equines, by such standards, might be thought of as "the last horses." And by the very same criterion, we, the members of *Homo sapiens,* represent "the last humans." In evolutionary terms, we are the sole surviving representatives of a dramatically unsuccessful lineage. We are, you might say, the last hangers-on in a long line of losers. Tough stuff to take. Science writer Brian Switek put it this way:

> There was never an "ascent of man," no matter how desperately we might wish for there to be, just as there has not been a "descent of man" into degeneracy from a noble ancestor. We are merely a shivering twig that is the last vestige of a richer family tree. Foolishly, we have taken our isolation to mean that we are the true victors in life's relentless race.[29]

This is, incidentally, why it is a mistake to speak of any particular fossil as being the critical "missing link" between us and our prehuman ancestors. The striking diversity that once characterized the hominid line surely confirms our evolutionary origins. However, it also complicates the task of sorting out the exact great-great-great-grandfathers and great-great-great-grandmothers whose progeny survived the pruning of that evolutionary tree to give rise to us, the last surviving humans.

DEMOTION

In the nineteenth century, some may have found the grim news of evolution to be tempered just a bit by artwork showing humans triumphant atop the tree of life. One such illustration was produced in 1874 by the German zoologist Ernst Haeckel, and featured in his book *Anthropogenie.* While Haeckel's drawing didn't show humans as the result of a straight-line climb up the ladder of life, it certainly did confirm our position at the very summit of the world of life. But Haeckel, of course, knew nothing of the wealth of prehuman fossils that would be found in the twentieth and twenty-first centuries. As a result, he placed humans ("Menschen") in a position of success and dominance supported by the highest branches of evolutionary diversification, much as Linnaeus had attempted to do earlier.

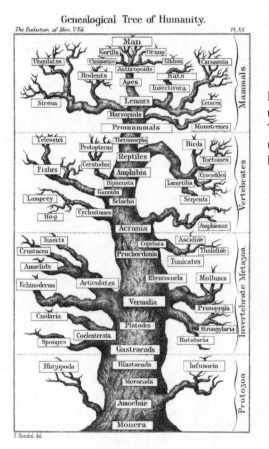

Figure 3-1: Haeckel's representation of the "Human Family Tree." The labels on this copy of the diagram have been translated from the original German.

While many people would still be content to imagine the tree of life in terms defined by Haeckel, the drawings we would make today are quite different. To a biologist, considering certain species as "higher" and certain others as "lower" is thought to be a mistake, since all living species are truly part of the same evolutionary process. That bacterium on the tip of your pencil is just as "evolved" as you are. It's found a quite different way to make a living, of course, but that's no reason to place it at the bottom of the tree of life or to place ourselves at its pinnacle. In that respect, a more accurate tree, emphasizing the evolutionary relationships among organisms, would look something like this:

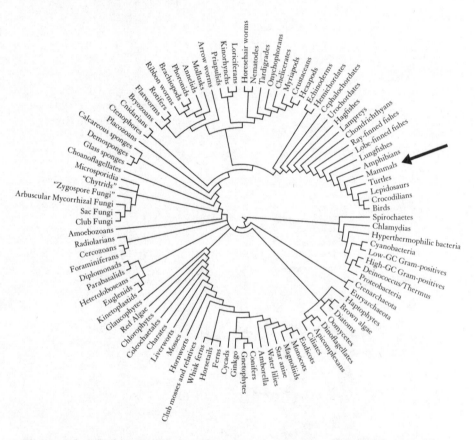

Figure 3-2: A highly simplified modern version of the tree of life, emphasizing the diversification of species from a common ancestor. In this radial version no species occupies a privileged position. *Courtesy of Dr. David Hillis*

This modern tree is still built around the concept of a single origin for living things, one that branches again and again to form a series of groups at its outer edge. The difference here is that no single living group occupies the top of the tree, nor is any such group placed at its bottom. Humans are included among mammals (the arrow in the diagram), and if the diagram is drawn in greater detail, mammals themselves are further divided into scores of coequal branches, each reaching to the very edge to represent a single living organism. As a result, there is no scientific reason to regard any branch as special, privileged, dominant, or preordained. Not even our own.

NOTHING SPECIAL

I am a cell biologist by training, and nearly all my research work has involved the use of microscopes, most especially the electron microscope. Developed at about the same time as television (and by many of the same scientists and engineers), the electron microscope is a marvelous tool that allows us to visualize tissues and cells almost at the molecular level.[30] It's an enormous privilege to be able to use such an instrument on a daily basis, and I never fail to be in awe of what the microscope reveals within the living cell.[31]

While most of my work has been with plant cells, I've done a fair amount of high-resolution microscopy on animal tissue, some of it human, and from that I can make an entirely mundane and obvious observation. There is nothing special about human cells. They are, in fact, indistinguishable from those of any other mammal. Given a set of micrographs, an experienced cell biologist could easily distinguish a liver cell from a muscle cell, or a pancreatic cell from one found in the nearby lining of the stomach. But try to tell a human kidney cell from the same cell type in a mouse, a bat, or a horse, and all bets are off. On a cellular level, life is one. And we humans, like other animals, are collections of cells.

Much the same is true at the molecular level. DNA is the genetic basis of all living cells. While there are indeed differences among species in the way DNA is replicated, genes are controlled, and development is regulated, these are more variations on a theme than fundamental differences. As a result, when biologists study the movement of proteins within cells or the mechanisms by which cells divide, they find nothing about human cells that distinguishes them from those of any other mammal, let alone those of any other vertebrate. If the key to life is to be found within the cell—and like all cell biologists, I think it is—then we have no particular need to favor human cells over those of any other complex animal.

These similarities are not limited to details at the microscopic level. They extend to qualities we tend to think of as uniquely human, including toolmaking, humor, and even moral sense. Writing to answer the question "Does evolution explain human nature?" Frans de Waal, an expert in primate behavior, pointed out that this is an unsettling idea for many:

> After all, the idea that we descend from long-armed, hairy creatures is only half of the message of evolutionary theory. The other half is continuity with all other life forms. We are animals not only in body but also in mind. This idea may prove harder to swallow.[32]

Indeed, for many it has proven impossible to swallow. But as de Waal points out, even our big brains are not quite as novel as we may think:

> If we look at our species without letting ourselves be blinded by the technological advances of the last few millennia, we see a creature of flesh and blood with a brain that, albeit three times larger than that of a chimpanzee, does not contain any new parts. Our intellect may be superior, but we have no basic wants or needs that cannot also be observed in our close relatives.[33]

Therein lies the problem. Our growing knowledge of comparative biology threatens our sense of uniqueness, shared across the histories of countless human cultures. One might wonder, however, if we can find solace in the grandeur of the evolutionary epic itself. Perhaps we should look at the past as prologue and view the sweep of natural history as leading inevitably to our emergence as the climax of life's drama on this small blue planet. If all of this was meant to lead to us, then ultimately our existence is the result of steady evolutionary progress, climbing up from ancient muck to the sublime perfection of the human creature. At the very least, we are here because we are the fittest, the best, and brightest, the sure and certain winners in a struggle for existence. Bracing stuff, if only it were true.

A WONDERFUL ACCIDENT?

In his book *Wonderful Life*,[34] the late Stephen Jay Gould argued that our presence on this planet was at least as much the stuff of luck as it was the product of natural selection. Typical for Gould, a paleontologist, he drew the attention of his readers to a particularly significant fossil formation, the Burgess Shale in British Columbia. The treasure trove of organisms captured in these rocks has long fascinated biologists, because they represent some of the very first complex animals found in the fossil record. The Burgess Shale animals date to the Cambrian period, roughly 530 million years ago, and they demonstrate a bewildering variety of structures, shapes, and body plans, only a few of which can be related to major groups of animals still with us today. The others have vanished into extinction.

Our own major group (which biologists call a *phylum*), the chordates, seems to be represented in the Cambrian by a single small organism known as *Pikaia*. If this tiny swimming animal, which could easily be mistaken for a worm, is indeed a chordate—and not all scientists agree that this is the case—its descendants include all of today's fish, amphibians, reptiles, birds, and mammals. But *Pikaia*'s body plan was just one among many in the Burgess formation. Some of the fossils can be recognized as antecedents of other phyla, including mollusks, arthropods, and annelids (the group that includes common earthworms). But many of the other Burgess organisms represent distinct phyla that did not survive past the Cambrian. If we are among the "winners," these lost phyla are surely the losers.

Now Gould asked a critical question. If we imagine the history of life as a videotape, rewind it back to the Cambrian, and then let it run again, would the outcome be the same? Would *Pikaia* once again survive, perhaps as a result of a superior body plan, while other groups from the Cambrian would once again disappear into extinction? Gould's answer to this was a firm no. There was nothing, he assured his readers, inherently fitter about any of the creatures of the Cambrian that preordained some of them for success. It was, in frank terms, a crapshoot as to which body plans would survive and give rise to the animal phyla we see today. We might have won the lottery, but no credit to us. It was, just as lotteries are, pretty much a random drawing.

But weren't specific trends nearly certain? Isn't it clear, for example, that mammals would once again survive the age of dinosaurs and diversify because of their inherent superiority? In a word—no. In fact, the rise of mammals provides one of the most telling examples of historical contingency in evolution. As Gould explains, the dinosaurs became extinct because of a completely random natural disaster—an asteroid striking earth:

> ... If the dinosaurs had not died in this event, they would probably still dominate the domain of large-bodied vertebrates, as they had for so long with such conspicuous success, and mammals would still be small creatures in the interstices of their world. . . . In an entirely literal sense, we owe our existence, as large and reasoning animals, to our lucky stars.[35]

If the asteroid Gould describes had missed the Earth by a few hundred miles, then the ecological niches we and other mammals fill would in all likelihood still be occupied by reptiles, and the world we know would never have come to be. What does this mean with respect to the question of human existence? That we owe our very existence not to some sort of innate superiority, but to happenstance.

> And so, if you wish to ask the question of the ages—why do humans exist?—a major part of the answer, touching those aspects of the issue that science can treat at all, must be: because *Pikaia* survived the Burgess decimation.[36]

We are here because a tiny, insignificant little critter managed to get out of the Cambrian period alive. And because of a lucky catastrophe that drove to extinction some of the most magnificent animals ever to walk the Earth.

In his book *The Accidental Species*, Henry Gee echoes this theme when he cites the principal mistake he believes we make when looking at human evolutionary history—the tendency to see ourselves as special or exceptional. As he notes, nearly every narrative that places us at the top of the evolutionary ladder cites the unique and exceptional nature of our species. But we are far from the only creature with unique and special properties:

> Giraffes are unique at doing what they do. So are bumble-bees, quokkas, binturongs, bougainvillea, begonias and bandicoots. Each species is unique by virtue of its own attributes—that's rather the point of being a species—and human beings are just one species among many. To posit humans as something extra-special in some qualitative way is called human exceptionalism, and this is invariably coloured by subjectivity. Of course we think we're special, because it's we who are awarding the prizes.[37]

Gee goes on to highlight many of the traits we often think of as exceptional to our species. But in each case he attributes these traits to a combination of factors, including chance, having little or nothing to do with the rise of humans to a position of dominance on the Earth. Bipedalism, for example, is often explained as an adaptation that allowed our ancestors to free up their arms and hands for toolmaking. But many other species make and use tools while still walking on all fours. And some of our distant relatives were indeed bipedal (Gee cites *Oreapithecus*, an extinct ape known from fossils in Sardinia and Tuscany) and yet never developed tools. So what is the connection? As Gee points out, bipedality is neither special nor uniquely human. He argues that each modern ape species has a distinctive mode of locomotion dictated by its own "very special evolutionary circumstances." Where bipedality evolved, "it was a trait as individual as any other variety of ape locomotion, not the first step in some progressive transformation between Ape and Angel."[38]

He links the remarkable increase in brain size in the human lineage (see Figure 2-1) not to toolmaking or even to natural selection for increased intelligence, but to dietary habits like the cooking of food, and to metabolic factors that made it possible to support so much high-maintenance nervous tissue. The fact that these elements combined in the human lineage, to Gee, is a matter of happenstance, not significance.

The same is true with language, which is highly developed in humans, but not unique to us, as we increasingly discover from studies of other creatures. And conscious self-recognition, such as the ability to recognize the image in a mirror as oneself? We can do that, of course, but so can other great apes, as well as elephants, dolphins, and crows.

The Accidental Species concludes with an afterword addressing the nearly universal human desire for a "Hollywood happy ending" to the evolutionary narrative. But such an ending is not to be.

> We human beings do like to tell stories, and the conventional picture of evolution as a stately and predictable procession with humanity at its head is just that—a story. As such, it speaks both to a profound misreading of Darwinian evolution, and to assumptions based on the fossil record that it cannot support, and never will.[39]

TELOS REGAINED

The Greeks had a word for the goal or purpose of an object or process. They called it *telos*. We might say the telos of a baseball bat is to hit a baseball, even though we may use it for other things. The telos of a pen is to write or draw, while the telos of a fertilized chicken egg is, of course, to develop into a chicken. Does evolution have a goal or purpose? Can we discern a *teleology* to evolution, something that would tell us of its ultimate goal and purpose?

According to the late Ernst Mayr, one of the twentieth century's greatest evolutionary theorists, the answer is no:

> Evolution itself was frequently considered a teleological process since it would lead to "improvement" or "progress. . . . [However] It is no longer a reasonable view when one fully appreciates the variational nature of Darwinian evolution, which has no ultimate goal and, so to speak, starts anew in every generation.[40]

If evolution itself has no ultimate goal, it certainly means that humans, as one of the products of evolution, were not its goal, either. Rather, as

Gee and Gould have pointed out, we are lucky survivors of a series of accidents and chance events, some cosmic, some mundane, but no more significant in the grand scheme of things than dandelions, beetles, or bacteria.

Where does this leave us? What can we now make of the mystery of our own existence, except to say that it is no longer a mystery? You might say that we once thought ourselves apart from nature, regarded, in nearly every culture, as handiwork of the gods. But then, our very anatomy betrayed us in our similarities to other animals. We were classified as primates, first among all animals to be sure, but animals nonetheless. Then we learned the most disturbing of all family secrets, namely that our ancestors, traced far into the past, were not human at all. Still, despite the embarrassment of parentage, we clung to the image of our kind as the final, ultimate victors in the war of life. But even that scant comfort would not last for long, as we learned the twisted and contingent pathways that were followed, without plan, design, or purpose, in bringing us to life. We know at last, in the manner of an unwanted child, that we were accidents. Not planned, not desired, not triumphant; merely another species ground out by the wheels of chance and uncertain fate.

Is this the only way to look at our position in the Cosmos, the reality of our situation and position on planet Earth?

We are indeed the result of evolution, but rather than saying we evolved *from* nature, it is far more accurate to say that we evolved *with* nature. The living world around us, the flora and fauna from which we take sustenance and often inspiration, is part of the same process of change. From that realization, we should take some genuine delight in knowing that we are the products of the natural world. This does not mean, as we have seen, that natural history follows a predetermined course or that the evolutionary pathway to human life was inevitable. Far from it. The specifics of natural history could have turned out quite differently, as Gould pointed out so persuasively.

However, one of the persistent qualities of life in all its forms is its drive to explore what biologists call "adaptive space." There are many ways an organism can survive and thrive, and over time, evolution tends to try them all. These include such obvious niches in adaptive space, as those occupied by powerful, meat-eating predators (lions, tigers, and bears) and by swift, elusive grazers and browsers (deer, elk, and gazelles). But there are many other, less obvious niches, occupied by such organisms as barnacles, mushrooms, slime mold, hummingbirds, and tapeworms. Before the great Cretaceous extinction that ended the reign of the dinosaurs, reptiles dominated many of the niches occupied today by mammals. Afterward, mammals adapted quickly to fill the voids left by the dinosaurs. The same was true after prior extinctions. Each time, organisms adapted through the evolutionary process to fill the space left by the extinctions, thus restoring a kind of balance to the many habitats and ecosystems of planet Earth.

One of those niches is the very one we occupy. We have come to fill it as a result of the evolutionary process, just like every other organism has. We are intelligent, self-aware, highly social, omnivorous, and capable of abstract, symbolic thought. That combination of qualities makes us newcomers to the living world. I certainly would not argue that our particular niche had been occupied in previous ages, only that such a niche has always been available. This time the evolutionary process found just the right creatures to fill it. In other words, our existence was dependent on, contingent on, and consistent with everything we know of the laws of nature. Our specific emergence as hairless bipedal intelligent primates was unpredictable, to be sure, but was not "random" in the sense that not just any creature or life-form could have appeared and flourished in "our" niche.

To a biologist, a niche is defined by the range of ecological conditions under which an organism can survive. At its most basic level, these include things such as temperature, moisture, available nutrients, and relationships to other organisms in the local environment. In that respect, we started out as unremarkable primates. We are an African species, and our physical adaptations fit ecological conditions similar to those of the other great apes of the continent. But for whatever reasons, our social adaptations, our abilities to communicate, and our hypertrophied central nervous systems set us apart from those biological cousins in an unprecedented way. These characteristics produced an exceptionally adaptable animal, one that could move well beyond the environmental web of its origin to create niches of its of own on every continent, in every environment. This adaptable animal could thrive under conditions that no African primate could ordinarily tolerate.

It is true that "we," meaning an organism exactly like *Homo sapiens*, might never have arisen and that natural selection has "no personality, no memory, no foresight, no end in view."[41] Yet, here we are, ready to wonder what we should make of our own fortuitous presence. Rather than seeing our existence as devalued by the process that brought us here, why shouldn't we instead take delight in every moment of consciousness? We should remind ourselves that the very *possibility* of human life was baked into the matter and energy of our universe from the very moment of its inception! Moreover, what distinguishes the human animal from all others is not the exact structure of its body, the details of its physiology, or its particular relationship to other branches on the evolutionary tree. Rather, it is our unique ability to learn, to reflect upon the nature of existence, and to make sense of the very process that produced us.

Why did all of this happen? Look again at Figure 3-2, showing the explosion of life, as Darwin would have put it, from a few forms into many. Consider how even today life continues to diversify, to innovate, and to explore new niches. Evolution continues to produce new species, novel capabilities, and unexpected adaptations. Viewed in this way, the emergence of a highly intelligent, profoundly social, superbly adaptable animal is nothing more than the result, perhaps to be expected, of an evolutionary process extending across time and space that has ceaselessly explored one niche after another.

Richard Dawkins once wrote that the universe as we know it displays "blind, pitiless indifference" to us and to life itself.[42] But he was wrong about that. The harsh universe he described is actually bursting with evolutionary possibilities—as our own part of that universe demonstrates. Remarkably, Dawkins's indifferent universe has given rise to creatures that contemplate life's value, beauty, and meaning daily. As if to highlight that point, those very creatures, having learned from evolution, now relentlessly search for life elsewhere in the universe.

What are the chances for success in that search for extraterrestrial life? No one can say for sure, but many years ago astronomer Frank Drake decided to frame the question in a systematic way. Drake fashioned an equation that listed a series of probabilities that, when multiplied together, would allow one to calculate the chances for life elsewhere in just our own galaxy, the Milky Way. This is the famous Drake equation:

$$N = R^* \cdot f_p \cdot n_e \cdot f_l \cdot f_i \cdot f_c \cdot L$$

Figure 3-3: The Drake equation attempts to calculate N, the number of civilizations in our galaxy, by multiplying seven factors together. In order, they are (R^*) the rate of formation of stars that could support life, the fraction of those stars (f_p) with planetary systems, the number of planets per solar system (n_e) suitable for life, the fraction of suitable planets on which life appears (f_l), the fraction of those on which intelligent life appears (f_i), the fraction of intelligent civilizations that communicate by releasing detectable signals into space (f_c), and the length of time that such civilizations release detectable signals (L).

Strictly speaking, we don't actually know any of the seven factors to a satisfying degree of accuracy. However, where astronomers once doubted the very existence of any true planets outside our solar system, it is now clear that planetary systems abound throughout our galaxy, a galaxy that is just one among 100 billion or more. Discoveries of Earth-like planets are now common, meaning that the conditions that gave rise to life on Earth may exist elsewhere, although we can only guess at the numbers of potential "elsewheres" as we gaze at even more distant galaxies and stars.

For all of its uncertainties, what the Drake equation does display, however, is optimism. Frank Drake clearly thought that since life had evolved at least once in our galaxy, surely there must be other places among the 200 billion stars of the Milky Way where it happened again. Not surprisingly, Drake was a colleague and friend of Carl Sagan, and together they fashioned the special plaque placed on the Pioneer spacecraft (launched in 1972 and 1973) designed to provide information about our species and planet to any extraterrestrial civilization that might encounter it. In a very direct way, Drake, Sagan, and the others who pioneered the search for extraterrestrial intelligence were making a statement about the nature of the universe we would do well to remember today. Life is not the result of supernatural forces or special exceptions to the laws of nature. Rather, life is a direct consequence of the laws of physics and chemistry, and the conditions of our universe. Life, in plain language, is to be expected, and intelligent life is as much the product of those conditions as any other feature of the living world.

Consider, if you will, a universe that explodes into existence in a burst of matter and energy. The fundamental forces and physical constants of that universe tread a delicate balance between too much and too little. They are not so powerful that the infant universe draws backward to collapse into itself, nor so feeble that its matter forever scatters and dissipates into the emptiness of expanding space. Instead, it coalesces into myriad clusters drawn together by those forces. Some become so large they press atoms together with the awesome power of gravity, producing enough heat and motion to ignite the fusion reactions of countless stars. The chemistry of these stars produces heavy elements that burst forth to gather into comets, moons, and planets. Before long, great clusters of these stars and planetary systems form into 100 billion galaxies, still receding into the ever-larger fabric of space.

On one of these planets, everything is right for something even more remarkable to take place. Chemical interactions among atoms formed in the furnaces of stars begin to produce self-sustaining reactions. Before long, those reactions have sealed themselves into compartments, and the compartments have begun to replicate. These living cells set off the process of evolution, and life not only spreads throughout this planet, it transforms it. Its atmosphere fills with highly reactive oxygen gas, released as living things capture the energy of starlight. Evolution produces microorganisms that dominate both land and sea. But it also experiments with living forms that crawl, climb, walk, and even fly through the air. And finally, in its almost limitless ability to explore the possibilities of matter interacting with matter, a creature arises with the ability to become aware of all that has gone before.

This creature discovers the laws and principles of its own existence; delves into the past, probing its own history and that of other living things; begins to explore and understand the vastness of the universe, the improbability of its own existence, and the intimacy of its kinship to all of life. Now, we may ask, how should that creature regard itself? Should it regard what has happened on this planet as a thing of no significance? Should it look at itself as just a temporary collection of atoms and molecules no different from any other collection of matter in the vastness of space and time? Or would it be justified—knowing that it gained its life, its presence, and its consciousness in an epic journey from the big bang to the present—to see its place in the universe as special, its emergence as an event of genuine significance, and its self-awareness as the triumphant realization of the universe's own potential? It is not a relic of religious superstition or an artifact of self-centeredness to think that the appearance of the human species matters for this planet and for the universe itself. Instead, it is a cold, rational appraisal of the fact that nothing like us has ever existed in the world of life we know.

Whether the consciousness, reason, and awareness displayed by human beings are the *telos*, the goal of the universe, I cannot say. But I can echo what astronomer Carl Sagan once did say, which is "We are a way for the cosmos to know itself." A material world that has finally produced a species capable of exploring and explaining its own existence has unquestionably reached a turning point in its history, and that is exactly where we stand today.

Chapter 4

Explaining It All

Rape.

It's an ugly word, difficult even to say. It describes one of the most reprehensible crimes we can imagine, and yet rape continues to happen, in every human society and at every social level. Why does rape occur? What motivates the rapist? Why, since all cultures express an aversion to the act of rape, do so many cultures nonetheless react in a way that shames the victim and sometimes excuses the rapist?

These are tough questions, and they play into long-standing arguments about the nature of rape. Did men *evolve* to be rapists? Is it a crime of sexual passion, or is it an act of power and dominance? Is it driven by a "rape culture" that might be subject to change or by innate biological drives over which neither society nor the rapist has full control?

In 1975, Susan Brownmiller famously argued that rape was "a crime not of lust, but of violence and power." In her book *Against Our Will: Men, Women and Rape,*[1] she claimed that rape was not merely a crime, but a tool of oppression that kept women in a social position subservient to men. As she wrote, "It [rape] is nothing more or less than a conscious process of intimidation by which *all men* keep *all women* in a state of fear."[2] This fear, according to Brownmiller, was the ultimate basis of sexual inequality in human society. The rapist, in her view, was not a criminal exception to the social order, but a key element in maintaining the basic injustice of that order. "Rather than society's aberrants or 'spoilers of purity,' men who commit rapes have served in effect as front-line masculine shock troops, terrorist guerillas in the longest sustained battle the world has ever known." To put it bluntly, as she often did, "all men benefit from rape" because it keeps women dependent upon men to protect them.

Brownmiller's book set off a raging argument about the nature of rape and clearly raised consciousness about the seriousness of the crime and its effects on women. It is widely credited with strengthening laws against rape and improving the ways in which the criminal justice system treats rape survivors. But Brownmiller's thesis on the causes of rape was not universally accepted and in fact was widely criticized by behavioral scientists. Especially controversial was her assertion that the male sex drive had little or nothing to do with rape. Invoking comparative biology, she wrote, "no zoologist, as far as I know, has ever observed that animals rape in their natural habitat, the wild."[3] By dismissing reproductive sex as a motivation for rape, Brownmiller's analysis placed the blame squarely on the nurture side of the nature-versus-nurture divide. Rape is not in our biology, it's in our upbringing and in patriarchal social institutions that, perhaps covertly, tolerate and even support the subjugation of women by the threat of rape.

Whatever the merits of Brownmiller's arguments, her assertion that rape does not occur among other animals was wrong. Behaviors that take the form of forcible sexual intercourse have been widely reported by naturalists.[4] More to the point, her exclusion of the male sex drive as a contributor to rape seemed arbitrary and politically motivated to many anthropologists.[5]

Her book was the work of a journalist long active in political causes, including the civil rights movement, and that was clear from the nature of her arguments. Comparing violence against women to the lynching of blacks in the American South, she saw both as intentional tools of a ruling class or gender determined to safeguard an arbitrary social order from which it drew benefit. As social constructions, both could and should change in the name of justice, in Brownmiller's view.

As Brownmiller was bringing the issue of rape to the forefront of discussions of social justice, a new movement was brewing in the biological sciences, the essence of which was the assertion that science for far too long had ignored the influence of evolutionary natural selection on human behavior. Promulgators of this movement believed it was high time to reevaluate everything the behavioral sciences told us about human nature in the cold, scientific light of Darwinian evolution. In 1980, two authors working from this model announced that Brownmiller had been mistaken. They had determined the actual causes of rape, and previous explanations for the motivations of rapists could now be discarded. They had found the answer, they asserted, because their analytical tools were based in *evolutionary science. A Natural History of Rape,*[6] written by biologist Randy Thornhill and anthropologist Craig Palmer, described the inclination to rape as a reproductive strategy, one that persisted in human populations today because it had been favored by natural selection. In the crudest possible terms, men possessed a natural urge to rape because evolution had put it there.

In their analysis, Thornhill and Palmer drew conclusions from a collection of statistics about rape. They noted that young males rape more frequently than older ones do, that young women of reproductive age are the most frequent targets, that rape occurs in all cultures, that laws to suppress it have had only limited effect, and that the psychological trauma of rape is greatest among women with the greatest reproductive potential. According to Thornhill and Palmer, these and other aspects of rape in human cultures could be explained in only one way: Rape is a behavior found in all societies, and, therefore, rape must be a result of the evolutionary history of the human species. Evolution can be used to explain the urge to rape in the same way it can be used to explain all other human behaviors, even the most mundane:

> When one is considering any feature of living things, whether evolution applies is never a question. The only legitimate question is how to apply evolutionary principles. This is the case for all human behaviors—even for such by-products as cosmetic surgery, the content of movies, legal systems, and fashion trends.[7]

As they made clear, neither author doubted that rape had to be a product of evolution, meaning that any nonbiological explanation for its prevalence, including Brownmiller's, could be dismissed out of hand. In their view, the only "legitimate scientific debate" about rape could be "whether rape is a result of rape-specific adaptation or a by-product of other adaptations."[8] Somewhat sarcastically, science writer Sharon Begley summarized their arguments this way:

> Back in the late Pleistocene epoch 100,000 years ago, men who carried rape genes had a reproductive and evolutionary edge over men who did not: they sired children not only with willing mates, but also with unwilling ones, allowing them to leave more off-spring (also carrying rape genes) who were similarly more likely to survive and reproduce, unto the nth generation. That would be us. And that is why we carry rape genes today. The family trees of prehistoric men lacking rape genes petered out.[9]

Begley went on to ridicule many of the conclusions advanced by Thornhill and Palmer, but her actual target was quite a bit larger. It was an entire field of science known as "evolutionary psychology" upon which they had based their analysis. Begley's somewhat caustic description of "evo psych" was hardly unique. In 1997, Michael Kinsley, serving as the moderator of a debate on William F. Buckley's television program *Firing Line*, explained that evolutionary psychology

> . . . applies the theory of evolution not just to physical attributes, but to a wide assortment of human behavior. Your decision to come to this debate tonight in this auditorium was dictated by pressures on our shared human ancestors generations ago. That's only a slight exaggeration of what the evolutionary psychologists believe.[10]

Well, maybe it was a bit more than a "slight" exaggeration, but evolutionary psychology does indeed attempt to explain human behavior by considering the present-day effects of selective pressures from the ancient past. In biological terms, today's technological civilization is something new, and if we really want to learn about our species, we've got to go back a bit. Thornhill and Barnes explained the importance of past conditions this way:

The difference between current and evolutionary historical environments is especially important to keep in mind when one is considering human behavioral adaptations. Today most humans live in environments that have evolutionarily novel components. . . . Therefore, human behavior is sometimes poorly adapted (in the evolutionary sense of the word) to current conditions.[11]

Rape, to use their language, might be one of those behavioral traits that is now "poorly adapted to current conditions" but served a useful purpose in the prehistoric past. That purpose, of course, was simply to ensure the successful propagation of the very genes that account for the rapist's behavior in the first place, no matter the damage to victims and survivors.

As you might expect, the reaction to Thornhill and Palmer's book was swift, contentious, and largely hostile. I will consider some of the objections to their work later in this chapter. For now, let's take their controversial conclusions as fair representations of how those in the field of evolutionary psychology have explained culture, behavior, and psychology as well as simple human likes and dislikes as adaptations shaped by Darwinian forces of natural selection. In so doing, it has placed the weight of science firmly on the "nature" side of the classic nature-versus-nurture argument. For one behavior after another, evolutionary psychology seems to say, "My genes made me do it." How did such an influential school of thought come to advocate the view that behavior and culture, likes and dislikes were all hardwired into our genetic destiny?

Believe it or not, you might say that it all began with ants.

OF ANTS AND MEN

I was lucky growing up, and one of the big reasons was the vacant lot right behind our house in New Jersey. It was as large as three regular house lots, which made it perfect for sports, especially for baseball. Life was good. A bunch of kids around my own age lived in the neighborhood, the New York Yankees were ascendant, and a couple of perfectly placed trees stood in as first and third base. In the summer, games took place just about every day, with makeshift rules and equipment, to be sure, but plenty of enthusiasm and even a little bit of skill as we became older and more practiced.

But we didn't play all the time, and that left me to daydream and explore, sometimes with a couple of friends, most times on my own. On warm days I'd wander into our outfield, shaded by large trees and bustling with all manner of bugs, watching as they flew or climbed or walked around us. My favorites were the ants. I noticed early on that there seemed to be separate colonies of red ants and black ants, and decided that these two groups surely ought to go to war with each other. When I found two nests close by, I'd use my hands to push the loose dirt into tiny ramparts, imitating the shapes of forts, and waiting for the inevitable conflicts to arise. Sometimes they did, as two ants would grasp each other, rolling around in a great tangle while I laughed. Mostly, however, they went about their business, ignoring my militaristic expectations in favor of gathering food and working mysteriously deep within their hives.

As I grew more curious about these little critters, I borrowed a magnifying glass to watch them running in and out of openings in the dirt that led to their nests. Through the lens, I could see armored bodies, jointed legs, and mouthparts grabbing the tiny bits of bread and sugar I sat out to attract them. The colonies seemed to act almost as single organisms, with workers gathering food for the common good and stashing it away even as they returned for more. More interesting still were the nests I could expose by lifting rocks. For an instant, the ants appeared shocked as their inner labyrinth of chambers and passageways was exposed to air and sunlight. But then, almost at once, they scrambled in response to this outrage, hustling to retrieve their precious larvae, little white bundles of life, and carry them to safety still deeper in the nest. The organized chaos was fascinating. The individual ants seemed to act in an almost random way, but their collective efforts did the job, and did it quickly. The organization, cooperation, and selfless dedication to the greater good of the colony were extraordinary.

What caused this, I wondered? Surely the ants must have realized they were in danger when I lifted up one of those rocks. And yet they didn't just run for their lives. Instead, they returned to the surface again and again to rescue their larvae, the colony's next generation, in selfless acts of heroism. Fascinating stuff. But I was a kid, and before long I was raking the infield and chalking baselines to get ready for our next game in that wonderful back lot. The ants were left alone, and for me the mysteries of their behavior were forgotten for another time and place.

Fortunately, others didn't dismiss this problem as easily as I had. Charles Darwin had already hit upon the idea that instinctive behaviors could be inherited, and therefore could be acted upon by natural selection. Therefore, the inherited behavior patterns found in any successful animal should favor the survival of the individual. In other words, every animal ought to be looking out for itself, taking care of number one. But that's not what the social insects do—individuals display a selfless, almost altruistic dedication to others, even to the point of sacrificing themselves for the survival of the group. How could this be? How could selfish, opportunistic evolution have favored individuals so dedicated to the survival of the group?

Darwin was also troubled by the fact that nearly all the members of an ant nest or a beehive were sterile females, incapable of reproduction. How could an inherited tendency for such remarkable social altruism have become common in a population when the individuals displaying the greatest extremes of altruism were themselves incapable of reproduction? While he did his best to offer an explanation, he admitted that these organisms did indeed present a "special difficulty" for his theory of evolution. And so it remained, well into the 1960s.[12]

The solution to this problem emerged gradually, as evolutionary theory became infused with the science of genetics and then with mathematical models of population growth and natural selection. For the first time, it became possible to study the effects of instinctive behaviors not just on the survival of individuals, but on the ultimate survival of the genes programming those very behaviors. In 1964, William D. Hamilton put all these elements together in two brilliant papers[13] analyzing how certain forms of altruistic, self-sacrificing behaviors might be favored by natural selection. Hamilton realized, as most others had, that a gene causing an individual to behave in a way that sacrificed its own well-being for the benefit of another would actually lower its chances of survival. But his great insight was that if that behavior were directed toward close relatives, such a gene might actually be helping copies of itself to survive. One's close relatives, including children, siblings, nieces, nephews, et cetera, are very likely to carry the same genes as you do. Therefore, a gene programming altruistic behavior within the family would actually, so to speak, be feathering its own nest in terms of natural selection. We call this provocative idea *kin selection*.

Hamilton went further, formulating this principle in simple quantitative terms. A particular altruistic behavior can be favored by natural selection if the benefit (B) to the recipient, multiplied by the degree of relationship (r) between the two individuals, is greater than the cost to the actor (C). Hamilton's rule takes this form:

$$Br > C$$

On one level, Hamilton's rule simply makes sense. Why, for example, should the mother of a litter of puppies devote her time, attention, and perhaps a thousand calories a day of milk to ensure that they survive? We might be tempted to say that the answer is obvious—they are *her* puppies, so, of course, she cares! But Hamilton's analysis enables us to look a little deeper into why the parent-offspring relationship should produce such altruism on the part of the mother. Caring for the litter exacts a considerable cost to mom (that's the C in the relationship), and in the wild may even reduce her own chances of survival. But that behavior produces an even greater benefit (that's the B) to the pups, since without it they would surely die. So, the reason we see behaviors such as maternal care in mammals is that the genes programming those behaviors are helping copies of themselves to survive. Since mother and pup have a genetic relationship of 0.5 (meaning that there's a 50 percent chance that any gene in mom is also found in an individual puppy), the very large benefit of the pups surviving far outweighs the cost to the mother.

In a way, Br > C explains why you would do almost anything to ensure the survival of your own children—and if you don't have children, it may help you to appreciate why your own parents doted over you incessantly. It may also explain why you care for your siblings, related by a factor of 0.5, much more than for your cousins, who are related by the much more distant figure of 0.125. The biologist J. B. S. Haldane once quipped that he would gladly lay down his life for "two brothers or eight cousins." I'll let you do the math.

Hamilton's second 1964 paper dealt directly with the social insects. In it, he pointed out that the peculiar nature of sex and reproduction in these critters practically guaranteed that extreme social behavior would evolve to produce colonies of highly cooperative workers. The key is something called *sex determination*. In humans and other mammals, sex is generally determined by a pair of so-called sex chromosomes known as the X and Y. If you inherit two X chromosomes (XX), you develop as a female. If you inherit an X and a Y (XY), you develop as a male. Birds are a bit different, and we call their sex chromosomes Z and W as a result, but sex determination is chromosomal there, too. Most insects follow similar patterns, and in fact it was in insects that sex chromosomes were first discovered by the pioneering biologist Nettie Stevens[14] early in the twentieth century.

But the social insects are different in that they don't have sex chromosomes. Instead, their sex is determined by how many complete sets of chromosomes an individual possesses. Those with two sets of chromosomes develop as females, while those with a single set develop as males. In most colonies, only the queen produces eggs. Eggs that are fertilized with sperm from a male develop as females, since they inherit two sets of chromosomes, one from each parent. The few eggs that are not fertilized develop as males, with a single set of chromosomes. This pattern of sex determination is known as *haplodiploidy*, a name derived from the terms biologists use to describe cells with single (*haploid*) or double (*diploid*) sets of chromosomes. We humans are diploid, by the way, since we inherit one set of chromosomes from each of our parents.

Haplodiploidy ensures that males will be rare in a colony, since most of the eggs a queen lays are fertilized and therefore develop as females. More important, it also produces surprisingly close genetic relationships between the members of a colony. Keep in mind that when the queen does mate with a male, known as a drone, all the sperm with which he fertilizes her eggs will be genetically identical. That's because he has just one set of genetic information. As a result, the offspring of that mating will be related to one another by a factor of 0.75, meaning they share 75 percent of their genetic information. This extremely close relationship, midway between full siblings (50 percent) and identical twins (100 percent), means that conditions are very favorable for the evolution of altruistic social behavior, since such behaviors will inevitably be directed toward very close relatives, which under haplodiploidy include nearly everyone else in the colony. It also means that the most effective way for a female worker bee to bring closely related offspring into the world is to sacrifice her own reproductive potential and do everything she can to help the queen produce more sisters for her. That's because if she flew off and mated on her own, those offspring would only be 50 percent related to her, while the queen's offspring are related to her and every other worker by 75 percent.

At a stroke, Hamilton had solved Darwin's special problem. The fact that worker ants and bees are sterile no long matters, because the concentration of reproductive potential in the queen ensures that the workers' collective efforts will, in fact, pass along their genes (including the ones for altruistic behaviors) as the queen lays more and more eggs. Haplodiploidy also, in effect, tips the scales of Hamilton's rule in the direction of altruism, since the "r" in that relationship (Br > C) is now larger (0.75). As a result, even very high-cost behaviors can now be favored by natural selection. We see exactly such behaviors in the remarkable willingness of worker ants and bees to sacrifice themselves to guard and protect the hive or nest.

All at once, the extreme social organization of this remarkable group of insects made sense. The queen is large and in charge, female workers labor selflessly for the collective good, and males are tolerated only when the queen needs to mate to replenish her supply of stored sperm. As I sometimes joke with my students, there's a great feminist science-fiction novel just waiting to be written around the theme of haplodiploidy!

As fascinating as the special case of the social insects might be, it's only natural to ask what's in it for us. What do we learn from studying them other than an extreme example of the power of evolution to shape behavior and form primitive societies? Roughly a decade after Hamilton's groundbreaking work, that question was taken up directly by a scientist who began his career by making the ants his life's work. His name was Edward O. Wilson.

Wilson's professional career took place largely at Harvard University, but he was educated at the University of Alabama, and one of his first scientific projects was a detailed survey of the ants of Alabama, his home state. Arriving at Harvard in the mid-1950s, he first concentrated on the classification of ant species, eventually developing interests in the details of ant societies and especially in the evolution of their social behavior. In 1971, he published *The Insect Societies*,[15] the most exhaustive analysis of the social organization of ants, bees, and wasps to date. I can only wonder how often Wilson had been asked, as scientists often are, why he chose to study the organisms he did. But I assume that question came up frequently in polite conversation, and I'd wager that over the years he found a way to answer it. In any event, he gave his readers such an answer on the very first page of *The Insect Societies*:

Why do we study these insects? Because, together with man, hummingbirds, and the bristlecone pine, they are among the great achievements of organic evolution. Their social organization—far less than man's, because of the feeble intellect and absence of culture, of course, but far greater in respect to cohesion, caste specialization, and individual altruism—is nonpareil.

These are excellent reasons for choosing ants as a field of study, and I'm sure that answers like this sufficed at faculty meetings and Cambridge cocktail parties. But Wilson's ultimate ambitions, certainly by the time he wrote his book's final pages, were far greater. The book's concluding three-page chapter bore the title "The Prospect for a Unified Sociobiology" and revealed his ambition to reach well beyond the world of bugs. Wilson wrote that he had been "increasingly impressed with the functional similarities between insect and vertebrate societies." He compared the social organization of organisms as distantly related as macaques and termites, marveling at their similarities, even while admitting to a bit of oversimplification in the comparison. But that, according to Wilson, was justifiable because " . . . it is out of such deliberate oversimplification that the beginnings of a general theory are made."

And what would that general theory be? As he made clear, it would encompass the biological basis of social behavior in both insect and noninsect societies. Thinking specifically of the vertebrates, he thought wistfully of how different their societies would be if they had followed patterns of organization like the social insects: "Had vertebrates gone the same route, the results would have been much more spectacular. The individual bird or mammal, possessing a brain vastly larger than an insect, might have been shaped into a comparatively better specialist." But vertebrates—humans included, he noted—had remained "chained" to a cycle of independent reproduction. That cycle "enhances freedom on the part of the individual at the expense of efficiency on the part of society." For what it's worth, we are stuck with the inefficiency that arises from individual freedom, but that does not mean that a general theory of social behavior cannot be formulated that would apply to vertebrates and even to humans. Wilson chose the name sociobiology to describe this general theory, and, having named it, he then proceeded to give birth to it.

A NEW SYNTHESIS

In 1975, Wilson published *Sociobiology: The New Synthesis*,[16] a monumental work that found its way not only onto the bookshelves of scientists but also onto coffee tables and into the public consciousness of social debates in the 1970s. *Sociobiology* was a large and expansive treatise on the evolution of social behavior that drew an audience far wider than Wilson's more specialized work on insects. Printed in a large two-column format and featuring striking wildlife illustrations done by Sarah Landry, it was an impressive work that defined the basic elements of Wilson's general theory, reviewing the mechanisms by which he believes societies are fashioned and maintained, and then demonstrating how those mechanisms work in a series of highly social species.

Wilson's survey began with microorganisms, moved along, naturally enough, to the social insects but then ventured into vertebrate sociobiology. He described how evolution had produced schooling behavior in fish, territoriality in frogs, dominance in reptiles, and nesting behavior in birds. Among the mammals, he considered herding behaviors in grazing animals, colony organization in prairie dogs, cooperative hunting in wolves, and the distinctive social traits of nonhuman primates such as baboons, gorillas, and chimpanzees. Had Wilson stopped there, I wonder if his work, impressive as it surely was, would have echoed much beyond the halls of academic science. But he did not stop there and included a final chapter on humans: "Man: From Sociobiology to Sociology."

Today, accustomed as we are to pronouncements from evolutionary psychology on the "reasons" for various human behaviors, it may be difficult to appreciate just how controversial Wilson's final chapter was when it appeared in the politically charged 1970s. Using the very same analytical tools he had applied to insect societies, Wilson searched across the diversity of human societies for characteristics common to all or most of them. Such characteristics, he reasoned, might well be the result of biological programming if they were found in diverse multiple societies whether primitive or advanced, agrarian or industrial, tribal or transnational. He then used sociobiological theory to account for the evolution of these shared characteristics, which included sexual pair bonding, the division of labor, ethical systems, religion, esthetics, and class conflict.

While it would be difficult to overstate the breadth of Wilson's ambitions for *Sociobiology,* so too it would be difficult to overstate the hostility of reaction against it. Two of his Harvard colleagues, Stephen Jay Gould and Richard Lewontin, immediately took issue with Wilson. They accused him of endorsing a theory of biological determinism that justified the prejudices and inequalities of the existing social order. Joining with other Boston-area scholars to form the Sociobiology Study Group, they produced a number of articles critical of sociobiology's attempt to assign a biological basis to human behavior. In November 1975, they published a letter in *The New York Review of Books*[17] linking Wilson's ideas to eugenics, racist immigration laws, and even to the gas chambers of Nazi Germany. In that letter, they accused Wilson of presenting a theory that attempted to "provide a genetic justification of the *status quo* and of existing privileges for certain groups according to class, race, or sex."

Given such rhetoric, a handful of progressive and left-wing groups took to vilifying Wilson on a personal level. In a famous incident at the 1978 meetings of the American Association for the Advancement of

Science, just as he had been introduced as a speaker, a protestor from the "International Committee Against Racism" dumped a pitcher of water on his head, chanting, "Wilson, you're all wet!"[18] Characteristically, Wilson dried himself off and delivered his lecture. The audience gave him a standing ovation—he was not the sort to be silenced. Neither was his insistence on applying the young discipline of sociobiology to that most complex and troublesome of social animals, *Homo sapiens*.

Later in 1978, Wilson doubled down on his critics by publishing *On Human Nature*, which he has described as the third book in a trilogy coming ever closer to unifying the "two cultures" of the social and biological sciences. His ambition was clear enough in the book's chapter titles, which included biological explanations of "Aggression," "Sex," "Altruism," and "Religion." The author's complete devotion to biological, rather than cultural, explanations of human behavior is clear in the book's opening pages:

> If the brain evolved by natural selection, even the capacities to select particular esthetic judgments and religious beliefs must have arisen by the same mechanistic process. They are either direct adaptations to past environments in which the ancestral human populations evolved or at most constructions thrown up secondarily by deeper, less visible activities that were once adaptive in this stricter, biological sense.[19]

In the lexicon of human sociobiology, there are only two possible sources of social behaviors, and both are explicable in terms of natural selection. Our choices—aesthetic, religious, political, and sexual—are either the *direct* result of successful adaptations to the environment or the *indirect* results of adaptations to past environments. Everything is adaptation, and evolution will ultimately explain even the most intimate and complex of human institutions, customs, and cultures. Every one of them.

As an example of how the logic of sociobiology can be applied to a complex and influential human institution, consider Wilson's treatment of religion in chapter 8 of *On Human Nature*. First, he points out that religion is both ubiquitous and ancient. Religions of every sort permeate the human cultures, present and past, and we find religious symbols and totems even in the ancient graves of Neanderthals. Next, religion persists, even into our modern, technological age where rationalists would have expected it to wither and die as an artifact of prescientific superstition. In fact, as Wilson himself notes, religion is probably an "ineradicable part of human nature."[20] Why is this the case? The answer must be, in evolutionary terms, because it enhances the fitness of its practitioners. Social groups that were able to coalesce around myth, ritual, and symbolism would have been more effective in gathering food, making war, and producing children than their nonreligious neighbors, and so they survived to pass along their genetic endowments to humans today. Conversely, religions that incorporated beliefs adverse to such goals, such as the Shakers, who practiced abstinence from sexual relations, were selected against and gradually withered into insignificance. The religions that survive and prosper today, including Islam, Christianity, and Buddhism, are those that most effectively matched the genetic predispositions hardwired into our biology by natural selection.

Notably, religious beliefs, symbols, and practices can be incorporated into distinctly nontheological social movements. In recent times, the cultish veneration of Communist icons such as the body of Lenin, devotion to symbols like the Nazi swastika or the hammer and sickle, and unquestioning obedience to a little red book containing the thoughts of Mao Zedong as a kind of sacred scripture testify to the universality of the religious impulse. By the logic of sociobiology, the persistence of this impulse implies a biological connection to genes that mediate conformity and group cohesion, providing the ultimate basis for religious beliefs. Because they conflict with the religious impulse, genes for selfish, individualistic behavior have declined even as genes favoring conformity and group identity have risen. These genetic trends explain the persistence of religious belief and, in Wilson's view, also provide the ultimate rationale for rejecting such belief:

> If this interpretation is correct, the final decisive edge enjoyed by scientific naturalism will come from its capacity to explain traditional religion, its chief competitor, as a wholly material phenomenon. Theology is not likely to survive as an independent intellectual discipline.[21]

Therefore, in a very direct way, sociobiology not only explains religious belief, it also explains it *away*. This is the message that sociobiology conveys: we are creatures of evolution, and our deepest desires, wants, needs, and thoughts are all explained by evolutionary biology. *On Human Nature* won the Pulitzer Prize for general nonfiction in 1979 and remains a classic, highly influential among evolutionary psychologists.

Returning to the question of rape, which Wilson did not directly address, we can nonetheless see in the pages of *Sociobiology* reasoning very similar to that employed by Thornhill and Palmer. After noting that human aggressiveness is innate in all societies, Wilson writes, "It pays males to be aggressive, hasty, fickle, and undiscriminating." While he is careful to note that sanctions against rape do exist in all societies, he clearly believes that such sanctions are indicative of an innate biological drive toward sexual aggression those sanctions are meant to control. This is exactly the sort of biological drive described in *A Natural History of Rape*, a drive that led that book's authors to conclude that the tendency to rape is an inherent part of the human behavioral genome.

EVOLUTIONARY PSYCHOLOGY

The term *sociobiology* is not widely used in academic circles today, partly because of the political furor that attended the publication of Wilson's trilogy in the 1970s[22] and partly because it has been displaced by a discipline with even more expansive ambitions, evolutionary psychology. Evolutionary psychology is a direct descendent of sociobiology in that it also employs analytical tools based on the evolutionary analysis of behavioral adaptations and addresses many of the same questions. However, the interests of evolutionary psychology range beyond the social behaviors to address all aspects of human psychology.

How much broader is its scope? David Sloan Wilson, a prominent evolutionary psychologist, put it this way:

> Historians will look back upon the 21st century as a time when the theory of evolution, confined largely to the biological sciences during the 20th century, expanded to include all human-related knowledge. As we approach the 1/6th mark of the 21st century,

this intellectual revolution is already in full swing. A sizeable community of scientists, scholars, journalists, and their readers has become fully comfortable with the statement "Nothing about X makes sense except in the light of evolution," where X can equal anthropology, art, culture, economics, history, politics, psychology, religion, and sociology, in addition to biology.[23]

In his view, which is widely shared, the field of evolutionary psychology can now lay claim to just about every discipline in the broad fields we call the social sciences and humanities. Indeed, given this assertion, it would seem that the only areas of study that might escape the dictates of evolution would be the physical sciences, whose independence from biology safely resides in the rigid dictates and mathematics of physical law, at least for now.

One can, of course, make a compelling argument that all of the humanities and social sciences are ultimately the products of human behavior. Therefore, if evolutionary psychology has reached a point where such behavior is calculable by a Darwinian rubric, then indeed nothing in any such field may make sense except in the context of evolution. Fair enough. But so sweeping an assertion depends almost completely on the ability of the biological sciences to connect the raw material of genetics to the subtleties of behavior. Has that connection indeed been made? Do we know enough of the link between genes and behavior to make such statements with any certainty? In some cases, yes, we do.

Even bacteria, to take organisms we often think of as "simple," exhibit social behavior. Clusters of certain bacteria cooperate to form biofilms, secreting polymers that bind them to one another and to surfaces from which they might otherwise be easily removed. Others practice a behavior known as *quorum sensing* in which they adjust their own patterns of gene expression when population density reaches a certain point. Quorum sensing is carried out by means of receptors that respond to signaling molecules released by other cells. When these signals reach a certain level, the receptors trigger responses within the cell that aid in both individual and group survival. The genetics of these behaviors in several species are well understood, right down to the molecular level of gene expression and signaling pathways.[24]

Genes directly linked to behavior have also been found in animals. One particularly well-known example was discovered long ago in *Drosophila*, the common fruit fly widely used as a model organism for genetics. Male fruit flies have an elaborate courtship dance they perform before females as part of the mating ritual. Several decades ago, a single mutation, now called *fruitless* for reasons that will be obvious, was discovered that dramatically altered this courtship behavior. Male flies with two copies of the *fruitless* gene perform their courtship dance in front of males as well as females and are receptive to courtship from other males, something males generally resist. Fruitless males are unable to mate with females, and as a result, their sexual behavior produces no offspring—hence the term *fruitless*.[25]

In the 1990s, publication of a paper [26] describing how this gene controlled both "male sexual behavior and sexual orientation" caused a bit of a public stir by virtue of its possible implications for the nature of human sexual orientation. It was featured in an article on "Homosexuality and Biology" in *The Atlantic* [27] that dealt extensively with the question of whether straight or gay sexual orientation was hardwired into the human genome. To be fair, the authors of the scientific paper were clear that their work applied only to insects and did not imply genetic causation for homosexual orientation in humans. Nonetheless, many in the popular press were quite willing to imply that if sexual orientation had been shown to be genetically determined in one species, the same might well be true in others, including humans.[28]

Scientifically, it's important to note that the actual operation of *fruitless* is much more complicated than a simple straight-gay on-off switch. The product of the gene is a protein that belongs to a family of proteins[29] known to switch whole sets of genes on and off. In addition, the effect of the *fruitless* mutant depends upon the state of several other genes with which it interacts in quite complex ways. Reflecting this complexity, more recent research studies have described *fruitless* as being just one part of "a web of regulatory interactions modulating courtship behavior."[30] Nonetheless, it is fair to say that *fruitless* is representative of a number of specific genes in the fly that have been shown to influence behavior. As such, they do indeed form part of the genetic raw material upon which evolutionary natural selection might affect social behavior.

While there is no known analog of the *fruitless* gene in humans, there do seem to be human equivalents to other fruit fly behavioral genes. One of these is *dunce*, a mutant that adversely affects learning and memory by producing defects in a specific biochemical signaling system. Significantly, a particular human disorder known as Rubinstein-Taybi syndrome seems to be linked to mutations in a similar signaling pathway, suggesting a common cellular basis for some of the complexities of learning and memory.[31]

These fruit flies provide a powerful model of how it is often possible to induce mutations, study their effects, and then trace their mechanisms downward through a maze of cellular and biochemical connections to the very genes subject to natural selection. In such cases, there is no argument about the link between such genes and the specific behaviors they control. But it is also true that the more complex behaviors underlying social behavior are difficult and often impossible to analyze in the same way. How, then, can we address the claims of evolutionary psychology to have uncovered the genuine causes of specific human behavior?

Let's look at a behavior that is common among humans—fear of snakes. I'll admit that the first time I came across a large snake during a camping trip, I absolutely froze in fear. Subsequently I learned to identify which species were harmful and which were not, and even managed to handle them without concern or anxiety. So, is fear of snakes a learned behavior based on an understanding that some are poisonous or an instinctive one bred into our evolved psychology? A number of studies suggest that it is the latter. Experiments with monkeys reared in the laboratory have shown that they acquire an intense fear of snakes merely when shown photos of snakes accompanied by photos of other monkeys displaying a fear response. Significantly, when those fear response photos were shown with photos of other objects, such as flowers and toy rabbits, they did not learn to fear those objects. Other studies have shown that humans react in similar ways, displaying physiological fear and dread far more intensely to images of snakes than to images of other objects (guns and speeding cars) that are actually more likely to cause them injury.[32] These experiments suggest that humans have a predisposition to acquire a fear of snakes. Combine this with the finding that related species seem to show the same predisposition and that fear of snakes would have been a useful adaptation, and you have the makings of a strong case for the evolutionary origins of the behavior. Similar arguments can be made for the evolution of an instinctive fear of spiders, of great heights, and of strangers, each of which would have presented a serious threat to the well-being of our human and prehuman ancestors.

What about a more subtle and complex behavior, such as romantic jealousy? As a recent review of the field by David M. Buss and his colleagues at the University of Texas noted,[33] evolutionary psychology hypothesized that jealousy in men and women would be triggered by different factors. In men, they reasoned, jealously would be triggered by the act of sexual infidelity itself. Why? Because with men the issue of paternity should be paramount in deciding whether to devote resources to the offspring of a romantic liaison. Sexual infidelity produces uncertainty about paternity, and therefore should be of greatest concern to a male. Among women, by contrast, provision of resources by the male is what should matter in terms of success for her offspring, and therefore a male's emotional infidelity, such as him falling in love with another woman, would be of greater concern than his occasional sexual infidelity. Buss and his colleagues note that while the research on this question is not unanimous, more than half a dozen studies using a range of methods have indeed supported the evolutionary psychology hypothesis explaining sex differences in the emotion of jealousy.

Incest avoidance is another case where science seems to have detected an instinctive behavior driven by evolutionary pressures. Inbreeding between close genetic relatives, such as brother and sister or parent and child, carries with it a high risk of genetic defects in offspring. This is because single copies of the same recessive disorders may be carried by both closely related potential parents, and therefore the children of such a union are at risk of inheriting two copies of such genes and manifesting such disorders. Anthropological studies have consistently shown that human cultures and societies have historically long-standing taboos against incest, all of which predate our modern genetic understanding of the dangers of inbreeding. Are such taboos learned behaviors resulting from experience, or are they instinctive, the result of evolutionary natural selection against sexual attraction between close relatives? In this case, an unintentional human experiment seems to have provided the answer.[34]

It was once a common practice in southeastern China for families to adopt young girls and raise them together with their own infant boys, intending them to marry each other as they grew older. Arthur P. Wolf, of Stanford University, took advantage of excellent household records on the island of Taiwan to inquire as to the long-term results of these "minor marriages." Since there was strong cultural approval of such unions, and since the girls and boys were unrelated to each other, there was certainly no social taboo against the practice. Nonetheless, Wolf's study showed that women who married a "childhood associate" bore fewer children than those who married an unrelated person they met outside the family. In addition, they were more likely to divorce and to have affairs with other men. Wolf characterized the apparent aversion against sexual partners known from childhood as "socially unnecessary but psychologically inevitable."

From the point of view of evolutionary psychology, the behavior exhibited in these marriages amounts to the exercise of a simple behavioral rule: *do not be attracted sexually to anyone you lived closely with as a very young child.* Individuals who possessed an inherited inclination toward this avoidance behavior were unlikely to produce defective offspring, while those lacking the inclination to avoid often did. Therefore, the effects of long-term natural selection produced an instinctive behavior that effectively avoided incest, even if it occasionally made "mistakes" such as those that affected the minor marriages of Taiwan, where the behavior was unnecessary.

What about beauty? Surely, good looks were not necessary for survival in the rough world of our Paleolithic ancestors, so what could evolution possibly tell us about what makes a pretty face? According to some researchers, quite a bit. A number of studies suggest that perceptions of beauty are strongly correlated with facial symmetry, so that faces whose left and right sides are nearly mirror images are considered especially attractive.[35] Why should symmetry matter so much? The explanations favored by researchers seem to center around what might be called the "good genes" hypothesis. A symmetrical face may indicate that physical development has taken place without interference from disease, injury, or parasites. In such cases, presumably the person with such a face would be an ideal mate in terms of health and fitness, and therefore would be more likely to be chosen as a mate by a member of the opposite sex. If a preference for facial symmetry really did result in healthier and more numerous offspring, presumably that preference would have been passed along by natural selection. The result today is the cross-cultural standard of facial beauty we recognize in movie stars and contestants in events such as the Miss World and Miss Universe competitions. Therefore beauty, that most sublime of attributes, resides not in the eye of the beholder, but in the harsh realities of evolutionary adaptation.

SO, WHAT'S THE PROBLEM?

If evolutionary psychology has excelled at anything since its inception just a few decades ago, it would have to be at generating headlines. In fact, it's not unreasonable to suggest that it may have a higher ratio of publicity to actual research than just about any other field of science. There is a certain logic to this, of course, because evolutionary psychology presumes to tell us about our innermost selves, our drives and motives, in a way that few other fields could hope to approach. While this ensures public interest in the field's latest findings, it carries with it an ever-present danger of overstatement and exaggeration. Consider shopping.

In 2009, the press was abuzz with "scientific" discussions of why women love to shop. A British newspaper article[36] had cited a study by David Holmes, of Manchester Metropolitan University, revealing "skills that were learnt as cavemen and women were now being used in shops." Specifically, those were the gathering skills of cavewomen developed in the ancient past to seek items that provided "sustenance, warmth, or comfort." Those skills have now led to "comfortable shopping malls and credit cards." The article was widely quoted and reprinted, although careful readers might have noticed the paper's admission that Holmes's study was commissioned by a nearby shopping center. The nature of the evidence underlying this study was not clear.

Later in the year, however, Daniel Kruger and Dreyson Byker, of the University of Michigan, jumped into the fray with an actual research paper[37] describing sex differences in shopping behavior and comparing these to the presumed division of labor between the sexes in hunter-gatherer societies. *Presumed* is the key word here. Their analysis began with a claim that it was "likely" in ancestral environments that men were hunters, "likely" that women were gatherers, and "likely" that hunting would have interfered with women's duties as caregivers to children. After referring to the division of labor among several contemporary hunter-gatherer societies, their research consisted of an online survey given to 467 college students at two American universities enrolled in an introductory psychology class. Students were asked to agree or disagree with questions like "I can often remember exactly where I entered a store," "I like shopping better when I am looking for something big," and "I feel good after I've been shopping."

The results of the survey, according to the authors, support their hypotheses. These include that women would use the location of specific objects to navigate their way around a store, which they associate with foraging. Men, according to them, scored higher on "skills" associated with hunting, such as agreement with the statement "When I am going shopping for a big item, I like to have help from friends." Enthusiastic press reports appeared under headlines like "Why Women Love to Shop," and one summarized the research this way:

> That's why, during this holiday season, you're likely to see a lot of men cooling their heels, and a lot of women shopping until they drop. It's mandated (or should we say human-dated) by the evolutionary progress that guided us out of the woods and into the mall.[38]

Really? Can we actually determine hardwired evolutionary reasons for human behavior by surveys of contemporary, well-educated college students correlated with what we think might be the "likely" social conditions of the prehistoric past? And do the answers on such questionnaires actually correlate with hunting or gathering "skills"? Whether I "feel good" after shopping might depend more on my current financial condition than my stereotypical sex role, and how I navigate around a store is surely more a consequence of how accustomed I am to shopping there rather than a predisposition to hunting. Lastly, nearly all the respondents to that online survey (91 percent born in the United States with an average age of nineteen) had grown up in a place in which the very gender roles the investigators sought to validate as genetic were part of an immensely powerful popular culture. How then could they, or anyone else, argue that such roles were remnants of our evolutionary past rather than the cultural constructions of present-day society?

To more than a few critics, the "science" described in these reports was laughable. Biochemistry blogger Larry Moran, from the University of Toronto, regarded these studies as indicative of the flaws of the entire field. Citing a made-up parody of the girls-love-to-shop research, he wrote:

> Miss Prism's story [the parody] is just as credible as the stories made up by evolutionary psychologists and that's a damning conclusion. It suggests that the entire field of evolutionary psychology is practically worthless as a science—maybe when we kick all of the evolutionary psychologists out of the universities, they can make a living by writing comedy.[39]

Despite evolutionary psychology's presence in universities and colleges, despite its journals and professional societies and annual meetings, critics of the field abound. As they point out, much of the work produced in the name of evolutionary psychology amounts to pure speculation and a series of "just so" stories that confirm the prejudgments of researchers about the origins of particular human behaviors. This is particularly true of studies that lack direct experimental evidence and instead rely on sets of statistics and anecdotes. In many respects, this is how Thornhill and Palmer assembled their controversial ideas on rape. They made no genetic studies, didn't attempt to search for a rape gene, didn't try to find common biochemical or molecular traits among known rapists. More to the point, they made no studies of human societies in which the actual "costs" and "benefits" of rape behavior might be evaluated in terms of evolutionary strategy. Had such studies been undertaken, it is fair to say the conclusions of Thornhill and Palmer might have been quite different.

One person who did ask such questions was anthropologist Kim Hill, whose area of study included the Ache people of Paraguay, a tribe still living as hunter-gatherers in ways supposedly characteristic of the ancestors of modern humans. Hill decided to test the assumption that rape would be a strategy leading to evolutionary success for males. He conducted an analysis[40] of the "costs" and "benefits" of rape as a reproductive strategy for a twenty-five-year-old man among the Ache tribe. The analytical model considered it a "benefit" each time the act of rape allowed the man to pass his genes along to offspring in the next generation. But the "costs" included the likelihood of the man being killed or injured in the attempted rape, the possibility of deadly retaliation by the victim or her family, and the possibility that the pregnant survivor of a rape would reject the child. When taken together, he found that the "costs" exceeded the "benefits" by at least a factor of 10. Summarizing these results, Hill stated that it wouldn't have made sense for men in the Pleistocene to rape as a reproductive strategy, so the argument that rape is "programmed into us doesn't hold up."[41]

If claims for the adaptive value of rape as a reproductive strategy are flawed, what are we to make of the many other claims of evolutionary psychology regarding human nature? Some would have us toss out every one of its results as Darwinian storytelling, often done in search of the scientific limelight. They might point to the case of Marc Hauser, a former professor at Harvard and once a leader in the field of primate behavior. Over the years, Hauser's work seemed to show that many behavioral traits we associate uniquely with humans are present in other primates such as tamarins, macaques, and Rhesus monkeys. These include self-identification in a mirror, recognition of syllable sounds, responding to human gestures, and even the possession of a distinct moral sense. All of this suggested that our "human" nature has a history long and deep among our ancestors and animal relatives. His work greatly strengthened the case for the use of evidence in evolutionary psychology based on comparisons to other primates. Unfortunately, some of Hauser's work could not be replicated. He was ultimately charged with and found guilty of scientific misconduct for the fabrication and misrepresentation of research data.[42] He resigned from Harvard, and his research career came to an end. Several years earlier Hauser had published *Moral Minds*,[43] a well-reviewed book arguing that evolution-ary psychology could explain the origins of human morality. Maybe it can. But in Hauser's case it would seem that attempts to account for what he called "a universal sense of right and wrong" might have led him to regard this sense as nothing more than an evolutionary artifact, one that he was free to ignore with respect to his own behavior.

Of course, I don't regard Hauser's case as typical of researchers in evolutionary psychology. Nor should a case or two of scientific dishon-esty condemn a whole field. Were that the case, there would be very little left to science, since fraud has been present in science since its in-ception. But I do think there is a tendency in the field to overreach, to jump quickly to conclusions that match preconceptions about the struc-ture of society and the nature of evolutionary pressures on the human species. I also think that critics are right to point out that the political and ideological views of the field's practitioners may well creep into their findings under the guise of objective science.

Putting this crudely, journalist and science blogger Annalee Ne-witz[44] wrote that as a group, evolutionary psychologists "believe that certain groups of people are inherently smarter than others. They write books about how rape is a natural part of human evolution. And now, with another scandal rocking the world of evolutionary psychology, we can officially welcome a new breed of mad scientists into the spotlight: evopsych douchebags." The scandals she referred to included Marc Hauser but also Jeffrey Miller, a University of New Mexico evolution-ary psychologist who pointedly tweeted that "obese PhD applicants" need not apply to his graduate program. If they couldn't stop eating, Dr. Miller reasoned, they wouldn't have the "willpower to do a dissertation."

Despite such nonsense, I think it's perfectly clear that evolutionary psychology, when properly done, can help us to understand certain as-pects of our own behavior. Consider this example of animal behavior from author and biologist George C. Williams, describing research done by anthropologist Sarah Hrdy:

> She studied a population of monkeys, Hanuman langurs, in Northern India. Their mating system is what biologists call harem polygyny: Dominant males have exclusive sexual access to

a group of adult females, as long as they can keep other males away. Sooner or later, a stronger male usurps the harem and the defeated one must join the ranks of celibate outcasts. The new male shows his love for his new wives by trying to kill their un-weaned infants. For each successful killing, a mother stops lac-tating and goes into estrous. . . . Deprived of her nursing baby, a female soon starts ovulating. She accepts the advances of her baby's murderer, and he becomes the father of her next child. . . . Do you still think God is good?[45]

Leaving "God" out of the picture for a moment, one may reason that male infanticide practiced within the harem is a straightforward prediction of evolutionary theory. By eliminating all currently nurs-ing infants, the new harem master would seem to ensure that his own genes, including ones that might produce or favor the infanticide behav-ior, would come to dominate the next generation. He would also ensure that his own efforts at care and protection of infants within the harem are not wasted on unrelated offspring. Therefore, such a behavior, in the harem social culture of these primates, would be highly favored by natural selection. As humans, can we apply our understanding of this behavior to our own societies? I believe the answer is yes, with one very strong qualifier.

Statistical studies of human infanticide have been carried out in many societies and among many cultural groups. Common to nearly all of these studies is a distant echo of the very behavior described by Williams. In one important Canadian study,[46] the likelihood that a pre-school child would be fatally beaten by a stepfather was found to be more than 120 times greater than the likelihood of such a child being fatally beaten by a genetic father. The Canadian study is not an outlier, since similar results have been found in the United States and other countries. These seem to be chilling statistics and might make one won-der why any woman with children would dare to remarry, knowing that her offspring would face the possibility of such lethal violence from her new husband.

Clearly, we can apply the same evolutionary argument that ex-plained the behavior of males in langur harems to human families with male stepparents. The comparison makes it clear why the rates of infanticide should be so different for biological and nonbiological fathers, and this is how evolutionary psychology can help us to understand an otherwise baffling statistic. By eliminating children not related to him, a stepfather can ensure that efforts to provide for his new family go only to his biological offspring, and not to the unrelated children of a previous mate of his new wife. Biology prevails.

But it turns out that the real question is not why evolutionary pressures are powerful enough to induce murder, but rather why they are so incredibly weak that in reality they almost never do. The actual rate of stepfather infanticide in the Canadian study was 321.6 per million. So, in fact, the frequency of such tragedies was fewer than 1 in 2,500. While every killing is tragic and every life is irreplaceable, the fact is that the killing of an infant by any father, biological or nonbiological, is extremely rare. One might fairly generalize that stepfathers, by a huge margin, love and care for their spouses' offspring effectively and are certainly not inclined toward violence directed at their stepchildren. If the drive to propagate one's genes, which resides at the theoretical heart of evolutionary psychology, is so powerful, we should ask what other forces exist that seem to have checked that drive so dramatically. What about human nature today has enabled us to largely escape the amoral behavioral chains of our evolutionary past? There must be another, even more powerful influence, acting on the behavior of stepfathers and everyone else, and I think we know what that is.

DARWIN'S MIRROR

Do you like what you see in the mirror? I think most of us would say "only sometimes."

Your answer might depend upon your age, the time of day, and how much effort you've spent getting ready for a close, critical look at your reflection. Mirrors aren't always kind, as we learn with passing years, but they tell a useful truth. That's why we have them around, even if we don't always like the image that stares back.

Now think of a larger mirror, a deeper one that not only shows our outward appearance, but also penetrates directly into the past, revealing our histories, our origins, our true character. You might call it Darwin's mirror, an image of the human condition cast according to the rules and techniques of evolutionary psychology. The picture of human nature it presents is not always flattering.

Our image in the mirror of evolutionary psychology is one shaped primarily by chance and necessity. We seek happiness, friendship, and love not because these things are worthwhile in their own right, but because the drives to achieve them are adaptations chiseled into the human psyche by our ancestors' struggle for existence. The mirror tells us that our thoughts, our values, and our motivations are not really our own and are certainly not what they seem to be. When we smile upon meeting a stranger, when we delight at a piece of music, when we kneel in prayer, we are actually following the dictates of a complex program running deep in the secret recesses of our nervous systems. That program has been shaped not by conscious reasons and desires, but by the brutal realities of life thousands of years ago that affected the survival of distant ancestors. As Steven Pinker describes it, "Our minds are adapted to the small foraging bands in which our family spent ninety-nine percent of its existence,"[47] and not to the conditions in which we find ourselves today. No wonder the world seems so confusing. No wonder the need for evolutionary psychologists to sort it out for us.

Therefore, it is only fair to ask how good a job they are doing. On one level, it is clear they are on to something. We are creatures of evolution, and if evolution shaped our bodies, right down to the particulars of bone, muscle, and tendon, then it clearly shaped our minds as well. But did it shape each and every aspect of the mind to the point where we can turn to evolution to understand every aspect of human thought, behavior, and ultimately every aspect of society and human culture? Here the answer is clearly no.

Let's return to the issue of physical beauty, which I had earlier presented as an example of how evolutionary analysis might explain the adaptive value of a preference for highly symmetrical faces. Any number of studies in the earlier days of evolutionary psychology showed that men's perception of an ideal female body shape was pretty much the same, regardless of nationality or culture. In general, guys seem to like female figures with a low waist-to-hip ratio, something you might call an hourglass figure. Therefore, this culturally invariant standard of beauty was widely taken as evidence favoring an adaptationist explanation for male attraction to such figures. However, other investigators began to wonder if this cross-cultural preference might have been influenced by the power of Western media, especially by exposure to repeated images of the "ideal" female body as possessed by movie and television stars. Therefore, they sought out a culturally isolated group for comparison, the Matsigenka people of southeastern Peru. The male preferences of this group differed "strikingly" from previous studies, which involved males from America and other Western-influenced cultures. The Matsigenka people preferred female figures with a much higher hip-to-waist ratio. In conclusion, the authors of the study suggested that "many 'cross-cultural' tests in evolutionary psychology may only have reflected the pervasiveness of western media."[48]

Similar findings have been reported for the perception of beauty in facial images. In one recent study,[49] 547 pairs of identical twins were asked to rate 200 faces in terms of attractiveness on a scale from 1 to 7. Groups of nonidentical twins and of unrelated individuals were then asked to do the same. If the perception of beauty were indeed a hardwired evolutionary instinct, one would expect the identical twins' scores to match each other's more closely than those of the other groups. They did not, and the researchers concluded, "individual aesthetic face preferences are truly shaped primarily by individual life experiences." In other words, evolution has not produced a universal standard of beauty that is passed along as a genetic preference. Indeed, one might well argue that if it had, then natural selection would have gradually refined human facial and body appearance ever closer to a single standard of absolute beauty and perfection. A brisk walk through any city or shopping mall will quickly confirm that this is not the case.

Whether evolutionary psychology can explain standards of beauty is one question, but there is a more pressing question regarding its most expansive claim, which is to make sense of everything human in light of evolution. Should I look into its mirror and see a collection of drives and mental modules inclining me to rape, to plunder, to assert personal dominance, and to find pleasure in images, sounds, and tastes reflecting nothing more than my Pleistocene heritage? Do I admit that

I am *not* an autonomous, whole, thinking person, but just a collection of adaptations? Do I regard even my values as adaptations, so that discussions of what is good or beautiful or true are really behavioral placeholders for the ancient drives that once elevated us, red in tooth and claw, just slightly above other animals? The problem is that at its most extreme, evolutionary psychology denies personhood and independence of thought, and turns the classic philosophical question of the good life on its head. All that is really left is the question of how can I best adjust my ancient urges to fit the changed circumstances of modern life.

Is that what evolution should mean to us?

Was David Sloan Wilson correct in describing this century as a time when evolution will expand "to include all human-related knowledge"? While the enthusiasts of evolutionary psychology would surely endorse his view that their discipline can now dominate art, culture, history, sociology, and religion, it seems clear that something is lacking. When we recall how evolutionary psychology successfully explained why stepfathers are much more likely than biological fathers to harm their children, we should also remember that it failed to explain why the actual incidence of such harm should be so low as to vanish into insignificance. If we invoke evolutionary psychology to explain the willingness of parents to sacrifice for their biological children, we should also ask ourselves why so many are willing to make even greater sacrifices for their unrelated adopted or foster children. And if we harken back to an imagined past when women gathered and men hunted, we should ask ourselves how we can be sure that such imaginings truly reflect selective conditions that continue to shape our behavior today. The claim that evolutionary psychology provides the full and ultimate explanation for everything human clearly falls short.

So, is it nature or nurture? The answer, of course, is yes. It is both. We are not blank slates. Our brains do not enter the world as empty memory banks ready to be programmed. We do indeed carry the marks of natural selection in our minds as well as on our bodies, but the question is what those marks might be and how deeply they influence behavior and culture. Here, evolutionary psychology can play an essential role in probing the heritage of human behavior and delineating the influences of genes and culture. Evolution has produced a set of broad behavioral instincts, but these do not lock us into a Darwinian cage that invalidates the idea of independent thought and reason by explaining every human tendency strictly as an adaptation.

John Cleese, the actor and humorist best known for his roles in television's *Monty Python* series, elegantly sends up the fundamental problem of evolutionary psychology in this monologue in which Cleese solemnly plays the role of "scientist"[50]

> Hello. *We scientists* have now located a gene which *we scientists* believe gives people the need to believe in God.
>
> In other words, this "God gene" releases chemicals into our body which create the impression that there is meaning in the universe. . . .
>
> Now, the discovery of this God gene is a big step forward in our quest to show that every bit of human behavior can be explained away mechanically, because we have now also located the gene which makes some people believe that every piece of human behavior can be explained away mechanically.

In attempting to explain just about everything else as a mechanistic product of natural selection, evolutionary psychology has overlooked one of the most important human activities of all. That would be science itself, and evolutionary psychology in particular.

Think of it this way. If we agree that our religious and ethical impulses are the products of natural selection (and I believe they are), why not extend that explanation to other human activities, such as science and mathematics? If we think of the ability to consider the moral consequences of our actions as a product of evolution, then surely the ability to construct mathematical systems is as well. While no one would argue that natural selection favored the evolution of a brain capable of multivariate calculus or Riemannian geometry, most of us do indeed possess such a brain. No one would claim that conditions in the Pleistocene favored those who could accurately calculate the value of *Pi* or determine the mass of the electron, but we have indeed proved ourselves capable of both. Yet not for a second would anyone argue that physics or mathematics can be "explained away" by evolutionary psychology. We hold, almost as a matter of faith, that science, if practiced rigorously, will give us a true and accurate picture of existence. The ability to do science is surely one of the greatest gifts the evolutionary process has bestowed on our species, but we do not dismiss it as a mere artifact of the process. Why then should we dismiss moral values as illusory or devalue art, music, and literature as mere throwbacks to our primitive origins?

One might say that evolutionary psychology has sought to explain every human activity except itself, and unfortunately it doesn't appreciate why this matters. A similar sort of confusion is apparent even in the careful writings of Edward O. Wilson. In his book *Consilience*, he makes this statement about religion:

> If history and science have taught us anything, it is that passion and desire are not the same as truth. The human mind evolved to believe in the gods. It did not evolve to believe in biology. Acceptance of the supernatural conveyed a great advantage throughout prehistory when the brain was evolving. Thus it is in sharp contrast to biology, which was developed as a product of the modern age and is not underwritten by genetic algorithms.[51]

But if we take not just religious faith but our analytical abilities as the result of evolution, as evolutionary psychology insists we must, the human mind most emphatically *did* evolve to believe in biology. In fact, it invented biology! In seeking an adaptive explanation for nearly every human activity, evolutionary psychology consistently overlooks the fact that it, too, is a human activity made possible by the evolutionary process. My point is not that evolutionary psychology is of no value. It has already helped us to understand the genetic glue that binds families and social groups, it has given us insight into some of our most basic fears and phobias, and ultimately it may help us learn how to organize social and political systems to maximize freedom and human potential.

However, the highly speculative nature of some of its methods, including references to the uncertain social structures of human prehistory and a willingness to infer a genetic basis for behavior before achieving a true understanding of the actual genetics of behavior lead almost unavoidably to conclusions that reach beyond what scientific data can support. As a result, the field has produced and continues to produce headline-making news that often defies common sense and, in many cases, fails the test of reproducibility. Claims about genes for rape, hunter-gatherer explanations for shopping, and just-so stories of adaptation are all part of this overreach.

Evolutionary adaptation has clearly shaped the human brain, and with it the basic mental framework by which we learn, interact, and seek to understand the world. But that framework is the beginning of human development, not its end. The findings of evolutionary psychology have been and will continue to be an important part of the exploration of human nature. But an obsession to reach deep into the past to find the hidden chains that bind us must be paired with the realization that we have evolved into the only creatures able to perceive those chains, to rise above them, and to achieve an independence of thought, action, and creativity that makes possible the great and lasting achievements of our species.

We are, then, creatures like no others, with extraordinary flexibility of behavior, powers of imagination, and above all, conscious self-awareness. That self-awareness has enabled us, *alone among living things*, to stand above the imperatives of survival and reproduction and seek to understand how we came to be.

Chapter 5

The Mind of a Primate

The marks of evolution can be found everywhere in the human body. Our limbs retain an internal pattern of bones inherited from the very first vertebrates to crawl upon the land, more than 300 million years ago. The three tiny bones (popularly known as the hammer, anvil, and stirrup) in our middle ears derive from jaw structures of reptilian ancestors. The routes traced by our facial nerves follow patterns first laid down in ancient fish. Even the genes that trigger development of our eyes and limbs and muscle groups are modified versions of the genetic plan we have inherited from nonhuman ancestors stretching back across endless centuries of deep time. And for many, that's the problem.

If evolution crafted our bodies by tinkering with ancient plans and designs, the same must also be true of our minds. While it may not be unsettling to know we throw a fastball with a modified version of the very same limb structure that other animals use to gallop, climb, or fly, can we say the same of our minds? Like those arms, our brains are modified versions of an organ that once served the needs of nonhuman primates as well as early mammals, amphibians, and even fish. Our brains are much larger, of course, but are they really different? Can we trust them to give us a true picture of reality? Or has evolution left us with a suspect organ forever tainted by the brutally haphazard process by which the human species emerged? If we truly possess the *mind* of a primate, can we even begin to make sense of the complex nature of existence?

PERCEPTION AND THE MATRIX

My dad loved photography. Cameras were his prize possessions, and he treated them like fine instruments. He could talk film speeds, F-stops, and depth of field like a master. He developed his own film and printed photographs in a corner of our basement that served as a darkroom once the sun went down. His subjects were everyone in our house, every trip, every family dinner, every sporting event, and every backyard picnic for years and years. That may be why my first image of the brain was something very much like a camera. I thought of my eyes as lenses, snapping pictures of the world and projecting them onto a screen inside my head. Deep within was a master controller, looking at one picture after another and in response, working the levers to move the arms and legs that carried that living camera around. The master's job was to analyze pictures and issue commands to the machinery of the body. As far as I knew, he did it logically and well because "he," after all, was me.

A cliché in biology is that we know the world only through the windows of our senses, and in my childhood conception of the brain, that was certainly true. Picture-perfect videos, sounds, scents, and tastes all made their way into an inner sanctum where the virtual "me" took it all in. Tucked into the command chair, I'd instruct my body to turn the pages of a book, grab a hot dog, or zip up my jacket, making one big decision after another. It all seemed so simple and so logical. To be sure, we do know the world through our senses, and we interpret that world using our brains. That much is certainly true. But what if those systems are mistaken? What if they were to give us an incorrect picture of the world, what if our perceptions are incomplete or, worse yet, deceitful?

Something like this was an essential plot element in *The Matrix*, a science-fiction movie depicting a not too distant future in which intelligent machines have taken over the world. Humans are exploited as sources of bioelectricity to power that world. To keep them in check, the machines have immobilized humans in nutrient pools and subverted their sensory inputs in a remarkable way. As a result, humans believe they are actively living and moving around in a normal world, but that world is nothing more than a "matrix," a pure illusion that serves only to keep humans pacified and immobile while their bodies power the machines. If you've seen the movie, you know that one individual, a computer hacker named Neo, spots repeated inconsistencies in the matrix and eventually escapes it to join a band of rebels fighting against the dominance of the machines.

For Neo, the choice to break out of the comforting world of the matrix is painful, both physically and psychologically. There is a new reality to come to grips with and a new struggle to join. Above all, there is the jarring disconnect between the real world and the matrix world with which his senses had been fed and programmed his entire life, right up to the moment of escape.

Many elements of *The Matrix* are implausible, but like all good science fiction, its plot contains an element of truth, a glimmer of possibility strong enough to make us reflect upon our own times and circumstances. In the case of *The Matrix*, the illusory world to which Neo is subjected is one in which the senses are deluded in a way that allowed the brain within to construct an artificial reality, one chosen by the machines to ensure their dominance. Significantly, while the machines program a series of sensory illusions, they leave the human brain free to interpret and analyze that sensory input. This is their undoing when Neo, with a little help from the outside world, breaks free and helps to spark a revolution. The movie's scientific message, if you will, is that even if that camera on the outside world is tricked, even if the images, sounds, and sensations presented to the brain are faulty, eventually the master will notice. The mind within will break the fog of illusions and will find its way to a genuine understanding of reality. It's a nice thought.

As I reflected on the movie, I began to consider something Charles Darwin had written long ago. This was not because of any allusions to evolutionary biology in the film, but because of a deeper and more troubling question that Darwin's own ideas led him to ask about the human mind. In 1881, the year before his death, Darwin wrote a letter expressing his conviction that the universe was not the result of mere chance. After offering that opinion, however, he stepped back, wondering if he, or any other human being, could really be trusted to consider such weighty issues:

> But then with me the horrid doubt always arises whether the convictions of man's mind, which has been developed from the mind of the lower animals, are of any value or at all trustworthy. Would any one trust in the convictions of a monkey's mind, if there are any convictions in such a mind?[1]

This is a problem far deeper than any difficulties we may face due to the imperfections or subversions of our senses. It hits at the very core of thought itself and plays upon the realization that our brains, as much as any other organs we possess, are the products of evolutionary natural selection. Since we do not trust mere animals to think and reason on the deeper issues of existence, how can we possibly trust ourselves?

Darwin was hardly alone in such doubts, and despite nearly a century and half of scientific progress since he wrote those words, his concerns have continued to resonate. Why, for example, do we have such difficulty in grasping the very small world of quantum physics, where particles can have the properties of a wave and where action at one place and time can influence a distant, seemingly unconnected event? Likewise, why does general relativity, which describes the curvature of space and the effects of gravity on time, seem to make so little sense?

Could it be, as Richard Dawkins once observed, that the problem is evolution? Our brains evolved, as he put it in a 2005 talk, to help us survive "in the Pleistocene of Africa" and not to grapple with quantum mechanics or the space-time continuum. Our everyday experience takes place in what one might call a "middle world," nestled in between the incomprehensibly small domains of quantum fluctuations and the unimaginably vast universe of black holes and the ever-twisting fabric of space-time. Therefore, in Dawkins's view, the human mind is not at all a reliable guide to what makes sense at scales above or below those that shaped the brains of our ancestors. As J. B. S. Haldane put it a bit earlier, "my own suspicion is that the universe is not only queerer than we suppose, but queerer than we can suppose."[2] Perhaps the universe itself is beyond our ken, and natural selection is the culprit.

OUR INNER PRIMATE

It would be one thing if the only problem posed by the possession of an "evolved" brain was that it made college physics an exceptionally tough course. But the rigors of natural selection should not be expected to produce perfection of mind any more than perfection of body. There are two reasons for this, the first of which is that evolution never gets to start from scratch. No organism living today was fashioned out of whole cloth as a perfect fit for the environment of its times. It's more useful to think of evolution as a tinkerer than as a designer. Evolution starts with existing structures and patterns, then crafts modifications here and there that pass the tests of natural selection, however clumsy and inelegant these modifications may be. Evolution never constructs an organ from the ground up, never lays out the circulatory system like a hydraulic engineer would, never completely rewires the nervous system to fit changing circumstances. Instead, it's pretty much stuck with the flaws and imperfections of what came before. It can do no better than to add some tissue here, change a shape there, or add a new capability by modifying an old one. This is *particularly* true of the brain, to which evolution, over time, has added a part here, expanded a part there, always forced to do so by modifying and adding to an imperfect preexisting architecture.

Paleontologist Neil Shubin, in his award-winning book *Your Inner Fish*,[3] explored this theme brilliantly with respect to the human body. Struck by the peculiarities of human anatomy, such as the strangely complex pathways of certain nerves and blood vessels, Shubin traced these to our evolutionary ancestry. Like all vertebrates, we begin development with an embryonic body plan very much like that of a simple fish, bilaterally symmetrical, with nerves and muscles and blood vessels laid out in a gridlike pattern. But then, as we grow in the womb, twists and turns emerge that give rise to limbs rather than fins, to ears and throats rather than gills, and to a complex skeleton that makes it possible to stand upright and walk on land. These variations on the fish body plan are what make us human, to be sure, but they are also the sources of many instances of incredibly poor engineering. The disks in our spines tend to rupture, the ligaments holding our knees together tear easily, and because our mouths are open to our lungs as well as our stomachs, we can choke and die on a piece of food. The inner, most primitive parts of our brains are sometimes referred to as "reptilian," but in truth that would be high praise. Reptiles appeared much later than the primitive fish whose body plans of muscle, nerve, and brain still form a starting point for the development of every human being. It is on this shaky framework that we build language, morals, art, and even science itself.

Another problem with the "evolved" brain relates to natural selection—which cares not a whit about perfection, reason, truth, or beauty. In the words of Harvard psychologist Steven Pinker, "the brain processes information, and thinking is a kind of computation."[4] Therefore, thinking, according to Pinker's model, is nothing more than information processing, and in our past the only criterion that mattered was survival and reproductive success under the conditions in which our species emerged. Our brains were honed by natural selection, which may have been perfectly capable of presenting to us a distorted reality, so long as that distortion helped us to prosper and survive. Therefore, while we can surely trust our brains to make decisions that optimize our chances in the struggle for existence (or we could back in the Pleistocene, at least), anything more is suspect. The brain houses the central hub of a physical survival machine, a modified primate organ, and, as Darwin wrote, who would trust the convictions of such a mind?

Finally, there is the material reality of the brain itself. In the past, we might have thought emotions were centered in the heart, that inspiration came from our ethereal souls, and that the brain did little more than cool the blood as it rose from the furnace of the body. However, as Carl Zimmer describes in his book, *Soul Made Flesh*,[5] the work of Thomas Willis and many others in the seventeenth and eighteenth centuries firmly established the brain as the physical seat of thought, sensation, and mind. The brain, like the rest of the body, is cellular. It is built up of billions of cells known as *neurons* that respond to and pass along impulses to other neurons. These cells form an enormous and complicated network that is linked to stimuli from the environment, and also to the body's *effectors*, which execute instructions from the system. When an intense flash of light causes the eye to blink, the stimulus of that light activates impulses that course through the system and quickly send commands to the muscles of the eyelid to pull it shut, protecting the visual system from damage. Seen this way, the brain is just another part of the survival machine we call the human body.

HARDWIRED

One of the first successes in understanding the nature of that machine came in 1781, when Luigi Galvani produced a twitch in the muscles of dissected frog legs by connecting them to a battery. Galvani immediately thought of nerves as living wires that relayed electrical commands from one part of the body to another. This discovery led his nephew, Giovanni Aldini, to put on a series of macabre demonstrations of the power of electricity to reanimate dead tissue. Making good use of a fresh corpse from the nearby gallows, Aldini produced muscle contractions that moved arms and legs, and even induced changes in facial muscles. Onlookers were so impressed by these facial movements that Aldini continued his experiments with the severed heads of decapitated criminals. As he perfected his techniques, he stimulated jaw movements, caused the eyes to open and blink, and produced expressions so lifelike that one observer was said to have died of fright after witnessing the spectacle.

The profound influence of these early experiments extended to the literary world as well, providing a scientific backdrop for Mary Shelley's gripping narrative, *Frankenstein*. If Prometheus was the titan who gave us fire, then Victor Frankenstein, the monster's creator, was, by her subtitle, *The Modern Prometheus*, infusing the fire of electricity to bring inanimate flesh back into life. While Shelley's novel broke literary and imaginative ground that has been well tread ever since, its scientific subtext has received somewhat less attention. Underlying the fantasy of dead tissue come to life is the revolutionary understanding that human life and thought are purely physical constructs, explicable in terms of the predictable interactions of matter and energy.

In the nervous system, the nexus of these interactions is the neuron. These are the cells that transmitted Galvani's bursts of electricity to muscles, that Aldini stimulated to impress his crowds, and that at this very moment are carrying the images of letters on this page deep within your brain. Neurons are, in some sense, very much like the wiring that Galvani imagined, but they are different in important ways as well. As with most of the body's cells, they produce and maintain an electrical potential, a voltage, across their cell membranes. There's nothing particularly special about neurons in that respect. Like other cells, they actively pump certain ions out while pumping other ions inside themselves. As a result, differences in the concentrations of certain ions build up across the cell membrane, particularly sodium ($Na+$), potassium ($K+$), and chloride ($Cl-$), and these produce a difference in electrical charge across the membrane. The interior of the cell is roughly 65 thousandths of a volt (65 millivolts) more negative than the outside. That may not seem like much (an ordinary flashlight battery produces a potential of 1,500 millivolts between its contacts), but across an ultrathin cell membrane, that's more than enough to drive a variety of cellular processes.

What most distinguishes neurons from other cells is what they do with this electrical potential. When stimulated, neurons can open and close ion channel proteins in their cell membranes in a way that allows currents of ions to flow almost like waves from one end of the cell to the other. In a copper wire, electrical charges flow directly through the highly conductive metal carrying a current at great speed. Neurons mimic wires by opening and closing their ion channels in a way that allows pulse-like impulses to run from one end of the cell to the other. These impulses move far more slowly than current does through a wire, but still fast enough to send electrical messages from the brain to the nether regions of the body in milliseconds.

When one of these electrical pulses, which neuroscientists call an *action potential*, reaches the end of a neuron, it triggers the release of chemicals known as *neurotransmitters* that can pass the impulse along to another cell. In this way, impulses can be passed along from one cell to the next. These connection points between neurons are known as *synapses*, from the ancient Greek word for "junction," the point at which two things are joined together. Communication within the nervous system therefore consists of a series of electrical events. These events are driven by ions moving across cell membranes. These movements generate a series of chemical events, in which neurotransmitter molecules are released as the result of nerve impulses. These molecules then cross synapses to trigger impulses in subsequent cells.

In this respect, the nervous system, including the brain, can first be understood as a network of closely linked cells that pass chemical and electrical messages between them. Input to the network comes from sensory neurons, cells that respond directly or indirectly to stimuli from the environment. Some of these cells (taste and smell receptors) respond to chemicals. Others (the retina) respond to visible light, to sound (the inner ear), or to physical pressure imposed on the cells themselves (touch receptors). The great Spanish microscopist and neuroscientist Santiago Ramón y Cajal was the first to establish the cellular nature of the system, formulating what is now known as the *neuron doctrine*. For that, he is regarded as the father of modern neuroscience, having set the stage for an appreciation of the workings of the brain as resulting from collective actions of an enormously complex network of individual cells. Neuroscience today is built on that foundation, as is our understanding of the brain itself.

ATOMS THAT THINK?

Digital computers were originally built around vacuum tubes, which operated as switches, opening or closing one electrical circuit according to the voltage applied by another circuit. The British quite properly called such tubes "valves," reflecting the ways in which these tubes could open or close a circuit, allowing input from one electrical source to control output from another. Groups of vacuum tubes could be wired together in a way that they formed logic circuits, allowing digitally encoded numbers to be added, subtracted, multiplied, or divided. Such circuits, now built almost entirely of solid-state microelectronic transistors, are at the heart of modern computers. The languages with which we code data and applications for such computers are designed for these digital circuits and provide, as any smartphone or laptop user knows, enormous power and flexibility.

Neurons, too, can be grouped to form circuits in which the input from one or several cells controls output from another. The reflex arcs first triggered by Galvani involved a small number of cells wired in a nearly linear series. An impulse applied to one cell triggered an action potential that cascaded along its axon, triggering an action potential in the next cell, and then the next until the final impulse was transferred chemically to a muscle cell, which then contracted in response.

Groups of neurons can act something like switches as well. In many cases, the neurotransmitter molecules released by a single neuron into a synapse are not sufficient to trigger an action potential in the next cell. Instead, several neurons must "fire" together at a single synapse to release a sufficient number of neurotransmitter molecules to trigger an impulse. This means that the "decision" to fire off an impulse is made at such a synapse by weighing the input from a group of cells and, if it is sufficient, switching on a new impulse in the neighboring cell. If not sufficient, that neighboring cell remains silent.

So, speaking simplistically, neurons indeed mimic the functions of transistors, and therefore, it's plausible to think of the brain as a huge collection of neural circuits in the same way that a modern computer is a similar collection of electrical circuits. The brain, to be sure, is far more complicated than any computer yet devised, and neurons are capable of far more sophisticated information processing than are transistors. But the image of brain as computer is compelling enough that many have been willing to use that model to explain the workings of the brain, including thought, as phenomena completely explicable in terms of biochemistry and cell biology.

The conclusions that might stem from this analysis are provocative. If we accept the computer metaphor as fact, our thoughts aren't really "thoughts," existing in some purely mental, almost magical realm. They are based in matter, and produced by a nervous system that manufactures thought by opening and closing molecular gates to allow ions to flow in and out of cells. Once we have discarded the notion there is anything particularly special about thought, the "mind" becomes merely a word we use to describe what the brain does. And, while we are at it, we might also lay waste to the notion of the "soul," and with it the many superstitions of the prescientific age. Doubtless there are many who would say, "Good riddance." But not everyone.

Well before the age of molecular neuroscience, the writer and theologian C. S. Lewis saw something like this coming, and worried deeply about what it might mean for the human conception of self:

> If minds are wholly dependent on brains, and brains on biochemistry, and biochemistry (in the long run) on the meaningless flux of the atoms, I cannot understand how the thought of those minds should have any more significance than the sound of the wind in the trees.[6]

Lewis, of course, was concerned about what this might mean for the significance of human thought and perhaps for the validity of the religious experience as well. While one might be tempted to dismiss Lewis merely as an apologist for Christian doctrine, consider the nearly identical concern expressed two decades earlier by the evolutionary biologist J. B. S. Haldane:

> If my mental processes are determined wholly by the motions of atoms in my brain, I have no reason to suppose that my beliefs are true . . . and hence I have no reason for supposing my brain to be composed of atoms.[7]

Haldane, clearly, was concerned about the validity of scientific thought and worried that a purely material view of brain function might render science itself suspect. Both were bothered by the notion that the mind, as resident in the organ we call the brain, was purely the stuff of atoms, molecules, and cells. Both sought, I am sure, a conception of the mind that affirmed its competence to explore and understand the material world, and one that validated human thought as serving something beyond the basic needs of our species to survive and procreate.

But both also feared that the scientific study of the human brain might dash those hopes.

To be sure, it is appropriate to call the brain a machine, specifically, a chemical machine. While it might seem absurd to argue that a subset of carbon atoms, namely the ones in our brain, can think, while other carbon atoms do no more than allow a pencil to darken a sheet of paper, this is indeed what a molecular picture of the brain implies. Consider the fact that when carbon and many other atoms are arranged in certain patterns they do indeed seem to influence our thoughts, even if they may not quite become part of them. Mood-altering drugs, many of which are deceptively simple in molecular terms, do exactly this. One example is lithium, which is used to treat bipolar disorder. The active principle in lithium-based drugs is the lithium ion itself (Li^+), which may displace larger ions such as potassium (K^+) and sodium (Na^+) at critical binding sites within the brain. While the details of lithium's actions are still not fully understood, there is no doubt that the ionized form of this atom can have profound effects on an individual's mental state.

Lithium is hardly alone. The shelves of pharmacies are stocked with chemicals that elevate mood, relieve anxiety, calm hyperactivity, and even dampen the urge to smoke. Being a child of the 1960s, I'm tempted to recall Grace Slick of the Jefferson Airplane singing, "One pill makes you larger, and one pill makes you small . . . " Given that era, one might also think of the guru of psychedelic drugs, Timothy Leary, urging young people to "tune in, turn on, and drop out," all with the assistance of profoundly mind-altering chemistry, a chemistry so powerful that it could actually change our perceptions of reality. If atoms and molecules can influence the brain so profoundly, making us happy or sad, anxious or relaxed, then perhaps, as Haldane worried, our thoughts are nothing more than the motions of atoms in our brains. We might as well get used to it.

GHOSTS AND DARK GLASSES

Classically, Western philosophy made a clear distinction between mind and matter. The division was articulated most clearly by René

Descartes, the seventeenth-century French mathematician and philosopher who devised the Cartesian coordinate system still used in analytical geometry. Descartes saw the body as a material object, subject to the laws of nature and operating according to the principles of engineering and physics. The mind and soul, however, could not be accounted for so simply, and therefore must be of an entirely different nature. While the body could influence the mind, and vice versa, the essence of the mind was nonetheless nonphysical. Today, many join Descartes in regarding the mind as separate from the body. If you recall the way I described my own childhood view of the brain at the beginning of this chapter, you'll see a clear—if not very sophisticated—example of Cartesian dualism.

Remember that childhood image of a little "me" sitting in the brain's control room? When I thought of my own brain, I might have imagined a single presence, a self-aware entity in the midst of all those tangled neurons conscious of itself and in control of thought and action. But twentieth-century philosopher Gilbert Ryle argued persuasively that that view is wrong. There is, as he put it, no "ghost in the machine," and he called the search for a mind distinct and apart from the body a philosophical *category error*, applying a term he had coined. Rather, he believed that thought is physical. It is part of the ordinary activity of the organ we call the brain, and since the workings of the brain are ultimately explicable in physical terms, then so is thought itself.

Even if all thought is physically based, surely we can trust our senses to present our minds with an accurate picture of reality, even if those minds are directly based in the material workings of the cell, as we now believe them to be. Seeing is believing, right? But as any cognitive psychologist will tell you, our senses are not at all as reliable as we generally think.

This is most easily demonstrated by means of optical illusions, pictorial representations that trick what is perhaps the most highly developed sense our species possesses, the sense of sight. I am particularly struck by a famous image of a solid cylinder sitting upon a checkerboard. A bright light illuminates the board from an angle, allowing the cylinder to cast a distinct shadow across the board.

Illusion crafted by Edward Adelson

Two squares of the board are labeled A and B, and the viewer is challenged to state which of the two is darker. The obvious answer seems to be square A, which is outside the shadow cast by the cylinder on square B. However, if one removes the squares bordering square B, as shown here, it becomes apparent that both A and B are exactly the same shade of gray:

Illusion crafted by Edward Adelson

This illusion, devised by Edward Adelson, of MIT, demonstrates one of the most important characteristics of our senses, namely, that they process information before presenting it to the conscious mind. Try as I might, even though I know very well that squares A and B are the exact same shade of gray, I cannot force myself to "see" that in the image. The reason is that our visual systems do not present images to the brain in the same way a digital camera might do it by faithfully recording each and every pixel. Even if we do not wish them to do so, sensory cells in the visual system determine the relative brightness of individual squares on the checkerboard by comparing each of them to their neighbors. In this way, they enhance the edges of objects. The cells also automatically recognize the soft edges of a shadow and adjust our perceptions of the squares within that shadow so they are perceived as a bit brighter than they actually are; this compensates for the effects of the shadow.

Putting it one way, if I wanted to gauge the exact shade of every square in the image accurately, I'd do much better looking at a numerical printout of each pixel gathered by a digital camera than by gazing at the image itself. In that sense, our visual system fails a test of accuracy. But in another way, it is actually quite sophisticated. Having evolved to look at scenes partly in sunlight and partly in shadow, it automatically compensates for the effects of those shadows by adjusting the signals it sends to our conscious brain to account for the effects of shading. In so doing, it produces what usually amounts to a more accurate perception of reality. In the example, however, our perception is tricked by the details of a carefully prepared synthetic image.

This sort of nonconscious processing is not confined to the sense of sight. Sound, touch, and even taste and smell can be tricked as well. Sensations are themselves physical events produced by other physical events, and this means we cannot guarantee that the ways in which our brain perceives or interprets these events will correspond to reality. Our only certainty may be that if our sensory perceptions were not honed by adherence to objective reality, they were indeed fine-tuned by the demands of natural selection. So, when we see at all, it is through a glass, darkly.

DARWIN'S KLUGE

I learned programming in the 1960s when computers were large, delicate machines tended by clusters of human beings who ministered to their every need. The computer staff at my university fetched and mounted magnetic tapes, threw toggles on a console to switch from one operating mode to another, loaded output papers of various types, and hand-sorted stacks of paper cards on which students like me had punched coded instructions in languages such as COBOL and Fortran. The principal computer at my college filled a large room, but its memory was limited and processing speed was laughably slow by today's standards. These limitations made it critical to design programs so elegant that they used only a few instructions and made minimal demands on the core memory of the university's one shared machine. Try as I might, my programs never even got close to "elegant." The word most often used by student consultants to describe my coding was "kluge." It wasn't a compliment. The word describes a solution to a problem that is clumsy, awkward, and very much inelegant. My kluge-like programs used too many lines of code, possessed needless instructions, and were often built around older lines of code that I had repurposed to carry out tasks for which they had not originally been designed. That made them slow and clumsy, and they often failed when they encountered input I had not anticipated. To be sure, eventually the programs worked, so I did pass the course. But they were no one's model of how to write code.

In a certain way, the "design" of the human brain is a kluge as well. *Kluge* is the title of a book by NYU psychologist Gary Marcus describing what he terms "The haphazard construction of the human mind."[8] Marcus cites a number of kluge-like features of the human body, including the backward wiring of the retina, which produces a blind spot in the field of vision. He explains these as the result of an evolutionary process that produced not a perfect body, but rather one that worked just well enough to survive the rigors of natural selection. Why should the mind be any different, he wonders, recounting the ways in which evolution has gradually constructed the human brain by modifying, again and again, the brains of its ancient ancestors?

In the early 1960s, neuroscientist Paul MacLean characterized the way the brain has changed over evolutionary time with his description of a *triune brain*. The modern structure of the brain, he argued, could be viewed as the result of three distinct periods during which newer brains were layered atop older ones, adding to the capabilities of older tissue, but not replacing it. The brain, therefore, was like a large, complex building around which new construction had been wrapped again and again, without ever displacing the original structures. As a result, ancient corridors and hallways had been preserved deep within the edifice, retaining their functions and defying attempts to modernize the building as a whole. The hindbrain is the oldest—controlling automatic functions like breathing and balance, and coordinating muscle groups. The midbrain, or limbic system, governs emotions, instincts, sexual behavior, and feelings of well-being. The forebrain, the most recently evolved portion, handles language, decision making, and higher cognitive functions. While the notion of a strictly triune brain with rigid boundaries between these three divisions is an oversimplification, the circuitry of the brain is in fact a poorly integrated mixture of the truly ancient, the very old, and the relatively new all working side by side.

The result of such haphazard construction is a mind prone to error, mental illness, forgetfulness, superstition, illogic, imprecision of language, and instances of very poor judgment. Evolution may have imbued our minds with sensations of pleasure that once helped distant ancestors to survive, but today these same sensations lead us to eat and drink excessively in ways that are actually detrimental to health. Behaviors that once helped us to secure mates and safeguard our offspring, today manifest themselves as anger, aggression, and pathological anxiety.

Therefore, we possess the brains of primates, evolved not for precision and accuracy, subtlety and morality but for the cold-blooded survival-centered demands of evolutionary natural selection. In many ways, this is a notion even more discomforting than the mere slander of ancestry implied by our descent, because it strikes at the very heart of humanism. It potentially undermines our values and mocks esthetics, morality, and even philosophy. Can the brain, a mere collection of cells and colliding molecules, reliably make sense of anything? Perhaps Darwin was on to something when he worried out loud about trusting the mind of a primate.

WALLACE'S CATHEDRAL

The eminent primatologist Frans de Waal says, "We are animals not only in body but also in mind."[9] This is surely true. But if we are going to examine the limitations of our evolutionary inheritance, it's also important to appreciate the ways in which many animals display the very sorts of intelligent behavior once attributed only to humans. In a recent book,[10] de Waal rolls out an impressive list of such behaviors such as toolmaking by chimpanzees and crows, social cooperation among trout, the geospatial senses of squirrels, facial recognition by sheep, and the prudent caching of food by scrub jays. In fact, one critic complained that de Waal had assembled so many examples that he became a bit repetitious. Maybe so, but surely that was his point. Intelligent behavior is not confined to our species, and in many ways certain animals actually qualify as more intelligent than humans. If you doubt that, try hopping into a small plane and finding your way, without navigational aids, to the mountains of Michoacán, a flight that monarch butterflies make every winter.

Darwin himself noted that the difference between human and animal intelligence is not one of kind, but of degree. Even the behaviors to which we assign the greatest moral significance are present in our animal cousins to a surprising degree. Chimpanzees and bonobos have clear social rules, protocols for settling disputes, and a well-developed sense of fairness. They weigh rewards and benefits in their interactions with other individuals, and, as de Waal points out, "Like us, they strive for power, enjoy sex, want security and affection, kill over territory, and value trust and cooperation."[11]

What this means, of course, is that human intelligence did not emerge at a single stroke, and many aspects of our behavior are also displayed by other animals. That much is clear. But it would be naïve in the extreme to maintain that the level of intelligence displayed by humans requires no further explanation. De Waal's book is a marvelous display of pure brilliance on the part of our animal cousins. But it's worth noting that the book was written by Dr. de Waal, not by any of the high-achieving animals he describes. Animals make beautiful music. But they do not compose symphonies or score and publish their music, much less convert it into digital format or (perhaps the most human quality of all) demand royalties for its use. There is something distinctly different about the human animal, and that something demands an accounting.

Earlier I mentioned the doubts that Alfred Russel Wallace harbored about the evolution of human mental capabilities. Wallace realized that even those he called "savages" possessed brains equal to those of British academics, and he despaired of finding an explanation in terms of natural selection, which he regarded as the sole driving force of evolutionary change. He would later turn to the Divine to explain the extraordinary capabilities of the human brain, believing that natural selection could not.

Wallace, however, failed to appreciate something that Darwin surely understood: *natural selection is not the sole mechanism driving evolution.* In the introduction to the first edition of *The Origin*, Darwin stated this plainly: "I am convinced that natural selection has been the main but not the exclusive means of modification." Curiously, no one seemed to notice, and Darwin became increasingly exasperated when his critics demanded that every attribute of a living organism be explicable in terms of selective advantage, using the many such attributes that were not as arguments against the theory. In the sixth and final edition of *The Origin*, Darwin complained that his argument for other "means of modification" had been "of no avail." Hoping to make himself clear for one last time, he wrote, "Great is the power of steady misrepresentation; but the history of science shows that fortunately this power does not long endure."[12]

In many ways, however, it has endured. Critics of evolution still demand "just so" stories of adaptation for each and every characteristic of an organism, and where such stories are lacking, they seize upon the opportunity to pronounce the theory a failure. But Darwin was right. Natural selection doesn't account for everything, as the elegant, classic 1979 essay by Stephen J. Gould and Richard C. Lewontin, "The Spandrels of San Marco," makes clear.[13] Gould and Lewontin call attention to one of the most prominent architectural elements of the great cathedral of San Marco in Venice, its strikingly beautiful spandrels. These triangular surfaces where the cathedral's dome-like rounded arches meet are tiled with glittering mosaics depicting biblical scenes that tell the stories of Christianity and its links to Old Testament events and prophets. One might think, as the authors point out, that the spandrels were constructed as the starting points of the surrounding architecture of the cathedral. But they were not. Rather, they are *byproducts*, formed by the intersecting multiple domes of the structure, meeting at right angles. Build a cathedral with multiple domes, and there will be spandrels at the points where they meet, *whether you want them or not.*

Gould and Lewontin used this architectural feature as a metaphor for characteristics that evolution may produce as byproducts of a sort without actually selecting for them. They describe, for example, such well-known processes as the *founder effect*, in which a very small founding population with a collection of genes different from the general population ends up in an isolated place. Certain characteristics may come to dominate this isolated population even when those characteristics are not particularly beneficial to survival. Other forces driving nonadaptive evolution change include mutations that might increase the growth rate of several tissues or organs at once. As a result, selection for an increase in size of one organ might also increase the sizes of others, leading to unexpected new properties. Gould and Lewontin favor a *pluralist* view of evolutionary change, one that recognizes many forces at work, in contradistinction to an *adaptationist* view that would see every human characteristic as the direct result of natural selection.

Let's consider again Wallace's wonderment at the intellectual capabilities of "primitive" people who, once they have been "civilized," are capable of most remarkable feats of intellect, artistry, and skill. None of these, in the words of Wallace himself, would have been advantageous in the primitive environments in which such "savages" lived and, presumably, had to face the rigors of natural selection. So, why do they have such godlike capabilities? In fact, why do we all? The human fossil record carries with it a remarkable record of evolutionary change. Some of these changes are quite subtle, such as the angle at which the pelvis articulates with the femur, the shape of certain bones in the wrist and ankle, and the arching of the lower jaw. But one change in particular is not subtle at all—the spectacular increase in the size of the human brain. In a geological instant, starting about three million years ago, the average volume of the brain increased from around 400 cc to its present value of 1,300 cc. No other organ in our body underwent so dramatic a change in size in such a short period.

Part of that change, of course, was genetic. Although we share a huge portion of our genetic endowment with the other great apes, researchers have been able to pinpoint a couple of changes in the human genome that may have helped to trigger an increase in brain size. One of these is a gene known as ARHGAP11B that was formed by the duplication and subsequent modification of another gene that is found in most mammals. ARHGAP11B, however, is unique to humans, and was formed after our lineage split from that of the chimpanzee. Transplanting this gene into mice produced much larger pools of brain stem cells, and enhanced growth in the neocortex, precisely the area of the brain where higher order functions such as language and reasoning take place.[14] Significantly, this gene is also found in Neanderthals and Denisovans, helping to account for their increased brain sizes as well.

What is less clear is how, and even whether, genetic changes like this might have been favored by natural selection. A brain with more neurons might be useful in some ways, but neurons are expensive in terms of energy (all those busy cells and their action potentials burn up a lot of calories), and this is one of the reasons humans have a higher metabolic rate than any of our great ape relatives.[15] We are able to maintain that rate only by means of an energy-rich diet made possible by cooking plant and animal foods to release their nutrients for efficient digestion. As a result, some anthropologists argue that the taming of fire and the introduction of cooked food may have supported the evolution of those big brains. This is a story the scientific details of which are still to be worked out. But we can surely see what happened to our brain as a result of that remarkable and remarkably swift increase in size, and ask whether this has something to do with human mental capacity.

The simplest explanation of increased human intellectual capacity is that bigger brains make for smarter animals. There is little doubt that this is true in a general sense. Size matters. The human brain contains approximately 86 billion neurons, and that's a lot of computing power. Famously, Stephen Jay Gould and Noam Chomsky have both claimed this striking increase in the size of the brain is what made language possible. In their view, natural selection was not directly responsible for the multiple physical and mental adaptations that make humans uniquely capable of speech. Rather, the brain grew large and complex as a result of evolutionary forces that placed a premium on raw intelligence. As mental complexity grew, a broad range of new capabilities became possible for body and mind, and speech was among them. In their view, language did not arise directly by means of Darwinian selection, but indirectly—an intellectual spandrel—as the brain increased in size due to other selective forces.

The spandrel view of the origins of language is enormously controversial among linguists and anthropologists, and many argue exactly the opposite point, namely, that language was indeed a specific human adaptation that was favored by natural selection.[16] But setting that issue aside, we should also ask whether there are any qualitative differences that distinguish the human brain from those of other primates, and, if so, how those differences might have arisen.

Two Harvard neuroscientists, Randy L. Buckner and Fenna M. Krienen, believe they have detected such differences, and they also have a pretty good idea of how they came about. They were spandrels!

The outer layer, or cortex, of the brain develops, as they point out in a 2013 article,[17] with a certain set of prespecified connections between nearby neurons. These connections carry impulses along well-defined pathways that are strengthened by learning and experience. However, because our brain is so large compared to those of our primate relatives, many of these local connections became disrupted as we evolved, leaving these cells to link to one another in wildly different patterns. This explosion of this *connectome* is most apparent in certain regions of the human cortex known as the associative cortices. These are regions that specialize in connecting with cells across the brain, and are especially important in decision making, memory retrieval, and self-reflection. In the language of neuroscience, Buckner and Krienen claim that as the brain increased in size, these neurons became "untethered" from the hierarchies that dictated the structure of our ancestors' smaller brains. As a result, they write, this rescaling of the human brain "shifts the predominant circuit organization from one primarily linked to sensory-motor hierarchies to a noncanonical form vital to human thought."[18] In other words, like a village growing haphazardly into a metropolis, rapid expansion may have radically disrupted the usual roads, highways, and traffic patterns that move information around the primate brain. In the jumble of growth and reorganization, new connections were forged that cut through the old structures. It is these connections, these new links, that account for the depth of consciousness and thought that distinguish us from other primates.

If they are right, the increase in human brain size did a lot more than just give us more neurons. It freed up cells in the associative areas of the cortex to make new and novel connections that produced many of the unique capabilities of the human brain. As science writer Carl Zimmer puts it, "The emergence of the human mind might not have been a result of a vast number of mutations that altered the fine structure of the brain. Instead, a simple increase in the growth of neurons could have untethered them from their evolutionary anchors, creating the opportunity for the human mind to emerge."[18] And, emerge it certainly did, to construct civilizations that would produce language, music, art, architecture, and science. It may be too early to claim we now understand the origins of these and other human creations. But we no longer need to puzzle, as Wallace did, over the extraordinary capabilities of "primitive" or "savage" peoples. All humans have these abilities, the results of an evolutionary process that gave us brains unique in the living world.

SPEARING CATFISH

As we've seen, although we depend upon it to know, experience, and understand the world, our brain is an imperfect organ. It's prone to error, it's easily tricked, and it carries the emotional and cognitive burdens of its evolutionary past. Given such a kluge of a tool, how can we make sense or be sure of anything?

I became acutely aware of the defects of my own cognitive systems many years ago while on a "survival" camping trip in the Scouts. Our instructor challenged us to get by in the woods for three days equipped with nothing more than a penknife, a few matches, and a plastic tarp. The challenge wasn't a surprise, since we'd been training for exactly this sort of thing. Building shelters was easy. The forest was thick with dead wood, rocks, and vines, and we quickly built lean-tos by lashing sticks together and weaving tarps back and forth between them to make them rainproof. A campfire came easily, too. We could have used the matches, but one of us was a champion bow-and-drill fire builder, and he soon showed off his skills to everyone's admiration. The hard problem was food. We'd been schooled in the available wild edible plants, but after the first day of gathering and chomping on roots, leaves, and berries, we were looking for something more substantial. Our campground was adjacent to the shallows of a lake, so I volunteered to try my hand at spearing the catfish that swam near the shore. I unfolded my penknife, lashed its handle securely into a sapling split at one end, and prepared to go fishing.

It wasn't long before a good-sized catfish swam into view right in front of me. I took a deep breath, tightened my grip on the makeshift spear, and stabbed right at the fish. Although my aim seemed to be true, I missed by a mile. Well, more accurately, by about four inches. Another fish came by, and I stabbed but missed again. Then again. The final time I threw my spear at a fish, and while it missed, the blade stuck itself into the muddy bottom of the lake. That's when I realized why I was missing. As I looked at the spear through the clear water, I saw that it appeared to be bent at an angle at the very point where it entered the water. When I pulled the spear out, I saw that this was an illusion. The shaft was perfectly straight. What had been happening, of course, was that the path of light was being bent by the change in refractive index between air and water. That's why the spear appeared bent at the interface, and it was also why I was missing the fish. They weren't where they seemed to be.

After a brief rest, and with considerable encouragement from my hungry friends, I tried again, this time compensating for refraction by aiming closer to my feet than the fish appeared to be. A couple of agonizingly close misses, and finally I speared one cleanly. Then another. Then, after a few more misses, a third. Success. Let me tell you, when you're really hungry, there's nothing quite like catfish grilled over an open fire.

What I had done, of course, was to figure out how my senses and perceptions were deceiving me and then correct them just enough to catch dinner for my fellow campers. Learning how to spear through the surface of the water was a pretty ordinary "discovery," and my correction was about as commonplace as could be. It was the sort of discovery that everyone who has ever tried to catch fish in this way has made for thousands of years. But it is also indicative of the ways in which our brain is capable of correcting itself, despite its many failings.

As human beings, we make such compensations all the time. Sometimes they're just as trivial as my spearfishing adjustment or getting used to the fact that objects in a car's side view mirror are closer than they appear. At other times, they are more profound. Evolution has given us a craving for salty and fatty foods, a preference that endangers the health of people in affluent societies. It's endowed us with a sex drive that leads to the propagation of the species, but also is the source of conflict that may have deadly consequences. It has given us a set of social signals like smiles, handshakes, and soothing language that may put us at ease in encounters with others but also enables us to be hoodwinked, swindled, and conned by those who exploit such gestures. Each of these reflects the difficulties of working with a brain shaped by the evolutionary process. But if dwelling upon these and other limitations of our mental systems shakes confidence in our own thoughts and perceptions, consider this. There is scarcely a human alive who is not fully aware of the ways in which these very human drives, urges, and emotions can lead us astray. Yes, the human brain is a faulty instrument. But the more remarkable finding has been well articulated in books like Gary Marcus's *Kluge*. The human brain is fully capable of consciously recognizing its faults and correcting for them.

This does not mean that each of us acts in a perfectly logical and rational way at all times. It does not mean that our evolved brains automatically correct themselves, or that we understand every mental flaw and misdirection to which we are driven by the brain's haphazard construction. But it surely does mean that we have the conscious capacity to identify our limitations and to adjust our understanding of reality by being aware of our perceptions, emotions, and impulses. Whether that capacity was selected for directly or arose as a spandrel we cannot say. But by the very act of analyzing our imperfections, we give testament to the possibility of overcoming them. That, it seems to me, is one of the things that makes the human animal unique.

AM I A COMPUTER?

In the past several years, I've read any number of articles either comparing the brain to a computer or claiming that it actually is one. These are provocative claims, and they reflect an increasing sense of confidence

that neuroscience is beginning to zero in on how the brain really functions at the most basic level. As we've seen, it's certainly possible to compare the workings of a neuron or, more properly, several neurons to the
way in which transistors regulate the flow of electrical current. Since
transistors are key elements in the central processing units of modern
computers, the comparison is a tempting one. We might, of course, accept the comparison as a metaphor, noting the ways in which computers
and animal brains both process inputs and produce outputs consisting
of electrical (or neural) impulses. Some, however, insist this is more than
a metaphor, that the brain actually is a computer, and that we should
understand it as such. Such thoughts lead to the expectation that at some
point in the future we might be able to download a person's memories,
experiences, capabilities, and thoughts into an inanimate computational
system. That system, that computer, would be for all intents and purposes, a mental duplicate of a living human being. This makes excellent
stuff for science fiction,[20] but is it a realistic expectation? And, more
important, if the brain is a computer, what does that tell us about the
nature of thought and the human mind?

One of the more provocative essays along these lines was written
for the online journal *Aeon*[21] by psychologist Robert Epstein. He argued
that the brain is not a computer and used a number of simple examples to explain why. One of these was a classroom exercise in which
he asked his students to draw an American dollar bill from memory.
The sketches he got, not surprisingly, were pretty rudimentary. Most
of them had poorly drawn portraits of George Washington in the
middle, the number "one" written out several times, and some even
had representations of symbols looking something like the seal of the
US Treasury. But none of them came even close to a faithful copy of
the currency. All his students had seen dollar bills countless times, of
course, and yet none of them could execute a drawing that was even
close to reality. On the other hand, "show" a dollar bill to a computer
by means of a scan or a digital photo, and a realistic image of the bill
will be stored in that computer virtually forever. When it is "recalled"
from computer memory and sent to a digital printer, the image will
outshine even the very best sketch that a human being can make from
the "memory" in our brain.

Epstein's point was this: The human brain does not process or store memories as we have come to think of them in the computer age. No matter how many times we have seen a dollar bill, there is no place in the brain where a pixel-by-pixel image of the bill is stored. There is, in his words, no "representation" of the bill stored in a "memory register" of the sorts used by even the most basic digital computer. Therefore, as he also wrote, the brain "does not process information, retrieve knowledge, or store memories." He goes on to argue that the brain is not a computer, and the analogy comparing it to one is not only wrong but also fundamentally misleading. We should get on with brain research without being "encumbered by unnecessary intellectual baggage" of the sort that comes from thinking of it as a computer.

Many people responded to Epstein's essay, but one of the most telling answers to his challenges came from my friend Jeffrey Shallit,[22] professor of computer science at the University of Waterloo. Shallit suggested that Epstein seemed to know very little about formal theories of computing, which are not restricted to the sorts of digital machines commonplace today. Invoking the foundational work of Alan Turing on computation, Shallit pointed out that in its own way, the brain does indeed process information, recall memories, create visual representations, act according to certain algorithms, and produce output. This makes the brain, according to the modern theory of computation, a computer. He summarizes his arguments like this:

> Humans can process information (we know this because humans can do basic tasks like addition and multiplication of integers). Humans can store information (we know this because I can remember my social security number and my birthdate). Things that both store information and process it are called (wait for it) computers.[23]

There is no question that Shallit and other critics of Epstein's essay were correct. Along those lines, it might be instructive to note that the original use of the word *computer* was to describe the work of humans involved in performing complex mathematical calculations. Computers were first devised to automate such work, so it certainly makes no sense to argue that the brains of those human "computers" were not well adapted to perform the very tasks that were programmed into the machines that replaced them.

My reaction to Epstein's essay was more along the lines of, "Tell me what you mean by a 'computer,' and I'll tell you whether the brain is one." If all one means by "computation" is a system that receives informational input from outside sources, processes it according to certain rules, and then produces an output, then, yes, the brain is indeed a "computer." But that's a pretty minimal definition. As a result, despite Epstein's theoretical mistakes, there is something to be learned from him and from others who have resisted the brain as computer analogy.[24]

The computers we use today store information in binary code. Each discrete element in its memory is either a one or a zero, depending on the position of a switch corresponding to that element. Their central processing units execute instructions in an orderly sequence of commands, recalling information from memory and writing information back into it. Even though a computer may have many such processing units acting together in a massively parallel system, the flow of action in each one involves following the commands of encoded programs, one at a time. New programs can be loaded into the hardware of the system, modifying how that processing occurs, which means that the entire system, hardware, software, and processing algorithms can be represented digitally.

If this were indeed how the brain worked, it would truly be a computer, not just in basic theoretical terms but also by means of a direct analogy to digital computers. But it is not. Neurons are living cells, not solid-state circuits. They grow and die, they change, and they act by means of ion flows across cell membranes, which are quite different from the movement of electrons through a conductive wire. Complicating things still further, about 60 percent of brain cells are not neurons. They are known as *glial cells* and were once thought to do little more than provide a kind of insulation around the neurons. However, we now know that *astrocytes*, one common type of glial cell, can regulate synapse formation, respond to neurotransmitters, reorganize neural circuits, and even function in memory formation. A single astrocyte in the mouse brain can make as many as 200 million connections with other cells, and human astrocytes may be even more highly connected. We know little about the details of these multiple connections and how they regulate the interactions that are key to brain function. As a result of these and many other cellular interactions, the brain can and does restructure itself. It is changed by experience, and it behaves more like a huge collective of interacting and competing elements than like the combination of well-defined modular components found in a digital computer. So, our brain is not a computer like the ones we have come to know and interact with in our daily lives.

This matters because it affects how we understand ourselves, and whether we think of ourselves as mere machines that act in predictable, programmable ways and could eventually be duplicated, if not replaced, by nonliving computer-based simulations of the human mind.

Could the human brain, yours or mine, someday be uploaded to a computer in a way that would duplicate our memories, thoughts, personalities, and individuality? Writing in the *New York Times*,[25] neuroscientist Kenneth D. Miller (no relation, despite the similarity in names), points out the mind-numbing complexity of the human brain in terms of cellular connections, biochemical diversity, and electrical activity, as well as its dynamic cell biology. He suggests that it would be "thousands or even millions of years" before we acquire the technological capacity to "upload" and re-create the mind of an individual, if ever. Even Gary Marcus, while maintaining that the brain is indeed a computer, admits, "it is unlikely that we will ever be able to directly connect the language of neurons and synapses to the diversity of human behavior, as many neuroscientists seem to hope. The chasm between brains and behavior is just too vast."[26]

This does not mean that we must retreat to magic or spiritualism to explain the workings of the human mind, something that many advocates of the brain-as-computer analogy seem to think is the only alternative to simplistic models of thought and perception. We do not need to postulate a ghost in the machine. But we do need to understand that today's computers, with their hardwiring and digital representations of real-world data, are not (and never will be) adequate models for the kind of computation carried out by the human brain. Genuine thought is not to be found in a coded algorithm written by humans that can compose music or write poetry, elegant and useful as such algorithms might be. No, genuine thought remains an elusive property of that emergent, delicate, self-sustaining, and self-modifying form of matter we call "life."

THOUGHTS OF A PRIMATE

Several election cycles ago, a leading presidential candidate sought to assure certain voting groups that he was indeed on their side when it came to the evil doctrine known as evolution. Striving to make this clear while not alienating voters who might think otherwise, he stated, "if anybody wants to believe that they are the descendants of a primate, they are certainly welcome to do it,"[27] but he, of course, did not. I'm sure there were plenty of biologists and biology students listening to that debate, and I wonder how many of them broke out into the same grin that I did when I heard this. In fact, that politician is a primate himself, as are his parents and his children, although I doubt many of his supporters were troubled by that little scientific inaccuracy. But primates we are, like it or not.

So what are we to make of ourselves as primates? On one level, the human brain is a suspect instrument. Its perceptions are clouded by the unconscious processing of sights and sounds even before we can become aware of them. Our innermost needs and drives are the remnants of evolutionary heritage and often lead to poor decisions and judgments. Even the sense that we have of ourselves as individuals can be explained as an artifact of neural processing. And even that is merely the rush of ions across cell membranes. As Darwin asks, can we trust the convictions of such a mind?

What if we *cannot*? It would mean, of course, that "thought" and "reason" are nothing more than a fine-sounding gloss we apply to the physical functions of an organ made up, like every other part of the body, merely of cells. If "mind" is what the brain does, and the brain is just another organ, why should we privilege its musings beyond those of any other tissue, or of our own brains beyond the brains of wolves or bats or worms? Why not sweep away philosophy, morality, aesthetics, and ethics as nothing more than a series of fantasy castles built upon the sands of our own self-deluding imaginations? Why not despair of knowing anything at all?

The curious thing about this line of reasoning is that ultimately it disqualifies every human conclusion about nature, including the scientific ones. If we cannot trust our brains because they are made up of atoms, Haldane wonders, then how can we be sure that atoms exist, because that is itself a conclusion of just such a brain. Therein lies the fatal self-contradiction of this particular negative. We must then turn to consider the positive, which is that we can indeed use those very human brains to probe the realities of existence.

Interestingly, as was the case with evolutionary psychology, those who would use brain science to "debunk" the reality of thought or the value of philosophy have a blind spot to the implications of such a program on their own activities. If the human mind is indeed incapable of making judgments as to what is true and authentic, then how could it be capable of knowing exactly that? What an odd piece of matter this must be—capable of producing the thought that we do not think at all, and then thinking it important to spread the word to other minds so they too may come to realize that their thoughts are illusions.

The same is true of morality, which is now routinely treated as an epiphenomenon of natural selection with no basis in reality other than social utility and group survival. Understandably, those who hold and propagate this view are eager to spread the word that the old cultural and religious systems of values and moralities are nothing more than arbitrary constructions, and, as Richard Dawkins has written, the universe is absent knowledge of good and evil. And yet they propagate this view precisely because of an absolute morality they hold themselves, of the validity and positive value of such a message on human existence, never seeming to be aware that a profound moral sense of the value of truth is indeed the driving force behind their efforts.

In truth, we humans have come to know and catalogue the limitations of our brains with increasing precision. Unconscious processing may sometimes cloud our perceptions, but our conscious recognition of that fact paradoxically often *sharpens* those perceptions. Evolutionary history may dictate the pleasures, anxieties, and prejudices of our minds, but we, distinct from other creatures, have discovered that history. Having done so, we are in the unique position of devising ways to overcome it. And if the findings of neuroscience seem to suggest a certain mechanical character to the brain itself, we should realize that the creative intelligence capable of generating those results came from just such brains. We can, as Gary Marcus suggests, "outwit our inner kluge."[28]

We certainly have, as E. O. Wilson writes, probed the brain to the point "where no particular site remains that can be reasonably supposed to harbor a nonphysical mind."[29] So much for the ghost within. But we should refrain from assuming that this in any way diminishes the wonder of the human mind.

Those who seek to deny "the mind/soul as spirit" by comparing the brain to a "lump of meat" are missing the point, Marilynne Robinson writes. She goes on to say, "If complex life is the marvel we say it is, quite possibly unique to this planet, then meat is, so to speak, that marvel in its incarnate form."[30]

It may confound understanding to say that that "meat"—the substance of a creature made up of atoms like carbon and nitrogen and phosphorous—can think. After all, does a carbon atom think? Can a nitrogen atom know love? Can oxygen get angry? But something happens to hydrogen when it becomes part of a living thing. Is the hydrogen atom that forms a tiny part of one of my genes alive? Is my DNA itself alive? Life, as we have come to understand it, is a collective property of many atoms and molecules. In very much the same way, thought is the collective property of millions, no, of billions of neurons. But the mere fact that mind has a physical basis does not devalue or undermine the validity of logic, the utility of science, or the reality of human thought and achievement.

What has happened on Earth during the evolution of our species may be exceptional, or it may be what eventually happens on any planet on which the physical conditions and material composition of its crust are right. But we are, as the late Carl Sagan loved to point out, not merely made of stardust. We are also a part of the universe that has become conscious and aware. We are the universe becoming conscious of itself. And the locus of that consciousness, the centerpiece of the Cosmos's self-awareness, is none other than the human brain.

Chapter 6

Consciousness

To steal a line from philosopher David Chalmers,[1] right now there's a movie playing in my head, and there's one in your head, too. It's a super-high-definition 3-D movie, complete with surround sound as well as hyper-realistic scents, touch-and-feel breezes, and even the sensation of sunlight on my skin. In addition, my movie has running commentary. I look at the wispy clouds overhead and wonder whether they are stratus or cirrus. A bird soars overhead, and when I spot the splash of color on its trailing feathers, a voice whispers, "red-tailed hawk." I turn to the pickup truck in my driveway, and the voice says, "Time for that oil change. And, oh yeah, we need some milk. Remember to get that on the way back." I'm immersed in the movie from morning till night. The movie is inside of me, of course, but it's outside of me, too. That's because the movie is not only *from* the outside world, for me it *is* the outside world. In fact, it seems so real that at times I forget it's a movie. What I'm describing, of course, is consciousness, mine and yours.

Many philosophers, and even a few biologists, see those movies as a problem. They doubt that the properties of ordinary matter can account for the power and subtlety of human conscious experience without reducing it to a set of chemical reactions lacking meaning or value.

Going further, they point out that evolution is a physical process, based on a purely material understanding of nature. Since everything human is supposedly accounted for by evolution, that means consciousness itself is a problem for evolution. Thoughts like these have led a prominent American philosopher to proclaim that the "neo-Darwinian conception of nature is almost certainly false." Is this so? Or is there a way to understand the phenomenon of consciousness as a byproduct of evolution that in fact emphasizes the meaning and value of human thought?

MY MOVIE

Let's start with the movie itself. There's a lot of green in my movie right now. That's because I'm writing from my backyard and it's a beautiful spring morning in New England. I'm surrounded by grass and bushes, and as I gaze at the woods nearby, everything I can see is green. But how should I describe my "green" to you? As leaves emerge in the springtime, they have a bright, almost dewy tint that brings a smile to my face. I know the color comes from chlorophyll, a remarkable pigment built around an intricate molecular ring that is very much like a similar ring in hemoglobin, the reddish protein that carries oxygen in our bloodstream. At the center of chlorophyll's ring is a magnesium atom. The alternating series of single and double chemical bonds surrounding it absorb light in the blue and red regions of the spectrum very well. That leaves the green wavelengths of light in the middle of the spectrum unabsorbed. Hence, the beautiful bright green of springtime in our backyard.

If you were here right now, I'm sure you'd agree. The trees are indeed green. But how can you be sure that you see "green" the same way I do? I suppose I could use words to try to explain the sensation I feel when I see that color. I might describe it as "warm and inviting," or maybe "vibrant and pleasant," but words like these could be used for other colors as well. I could say that leaves are green and blades of grass are green, and I could even pick up a crayon of that color to show you what green looks like. "See, this is green!" I could point out that green is the "go" color on a traffic stoplight. But try as I might, I really could not be sure that you feel the same sensation as I do when I see green. I can't be sure that your movie is exactly like mine.

A similar challenge would be to describe the sensation of green to a person blind since birth. I could tell them which objects were green and which were red, blue, or yellow, but that would make little sense to someone who has never actually "seen" any of these colors. If I tried to describe the sensation of green, I'd be stuck with the same problem I have in describing it to you. Imagine, for example, that for you the color green produced the same sensation as red does to me. Every tree, every leaf would produce the sensation of red, but since birth you would have associated that sensation with the word "green." You and I would have, quite literally, no words to convey to each other the intense subjective experience of seeing these colors. More on that in just a bit.

Where does my green "movie" come from? Who directs, who produces the movies we all experience every day? The simple answer is they are products of a nervous system that manages input from tens of thousands of nerves and presents them to us as a set of experiences that are, well, movie-like. But this consciousness is much more than a simple viewing space for messages from our senses with all input on the screen, all the time. We, unlike a camera, can choose to become more aware of some things and to ignore others. When I focus on a difficult task, like driving a nail into a board or opening a combination lock, other sights and sounds almost disappear from my consciousness. It took me weeks of very conscious concentration to learn how to shift the gears and manipulate the clutch of a car with a standard transmission. But when I drive through traffic today, I rarely think of the coordinated actions of gas pedal, clutch, and shift lever. These happen automatically, even though I can bring them back under direct, deliberate control any time I like.

Sometimes the movie of consciousness plays out in unexpected ways. When I walk into a crowded room, I might scan it, looking for a familiar face, especially if I expect people I know to be there. When I do so, I am consciously activating memories and asking my brain to see if any of the faces before me match those memories. When they do, my plan is to walk up and say hello. However, sometimes this sort of facial recognition happens automatically, almost to the point of embarrassment. Several years ago, as I made my way through a crowd, a familiar face suddenly came into view. "Hello," I said, showing a warm smile of recognition. "Great to see you!" I extended my hand in greeting, only to see her flash a perfunctory grin and turn quickly away. A few minutes later I realized that I didn't know this person at all. She was a minor actress in a TV show I had watched from time to time. We'd never met, of course, but my personal system of facial recognition had been switched on when I saw her, and I responded almost as if she were an old friend. Understandably, she wasn't expecting this display of familiarity from a total stranger.

My thoughts and feelings are part of this movie, too. I can ask myself what I want to make for dinner, whether I should turn left or right at the stop sign, and if I should swing at the next pitch in a softball game. Some things give me pleasure and put me in a great mood for the rest of the day. Other things, like tragedies in the news or failure at a task I know I should have accomplished, get me down and affect the way I look at everything else. I like country music as well as rock, and the right sequence of guitar notes will always make me smile, whether it comes from Bruce Springsteen or Alan Jackson. At various times I'm bored, excited, depressed, intensely curious, joyful, or circumspect. These moods and feelings go well beyond the processing and presentation of sensory information, affecting how I interact with the world and with others around me. They are part of the inner "me" that is not only watching my movie but is also controlling my part of it, making decisions, and being conscious of my presence as part of that movie.

This is the phenomenon of consciousness. Of all the things that take place in the brain, it is the one process with which everyone is intimately acquainted. It's easy to experience, because consciousness is experience itself. Despite that, it is one of the most difficult to explain.

WHO MADE THE MOVIE?

Let's assume the obvious, which is that human consciousness is a product of the workings of our nervous system as it interacts with the rest of the body and with the outside world. In other words, that consciousness is a physiological function in the broadest possible sense. What that means, of course, is that consciousness, like every other human characteristic, is a product of evolution. Therefore, it is perfectly reasonable to ask how evolution could have produced the movie that runs within our heads.

We all experience consciousness in the most direct and intimate way. So, we might expect neuroscience to be well on its way to explaining the nature of consciousness and accounting for its emergence from our evolutionary past. If you are at all familiar with the literature on consciousness, however, you know that this is not even remotely the case. In fact, the apparent ease with which we describe consciousness may be part of the problem.

In 2016, *New York Times* science writer George Johnson attended a Science of Consciousness conference in Tucson, Arizona. As he scrambled from one disconnected session to the next, he realized that there was a lot of talk but not much enlightenment:

> In even the wildest presentations, one could sense a longing for an answer to the question of consciousness, a fuller accounting of what we are and how we fit into the cosmic machinery. For all of the effort, the goal of providing a compelling explanation—one so clear it would make your head go bing—seemed as remote as ever.[2]

Books about consciousness have proliferated in recent years, some seeking to pin down this elusive process, others claiming that the mystery of consciousness will remain forever beyond our scientific grasp. Amid this emerged the claim, remarkable to me at least, that evolution as we know it could not have produced the phenomenon of consciousness. Philosopher Thomas Nagel made this argument in his 2012 book *Mind and Cosmos,* subtitled *Why the Materialist Neo-Darwinian Conception of Nature Is Almost Certainly False.*[3] Nagel is a highly respected philosopher of science, and he was surely on firm ground to point out that consciousness remains an unsolved scientific problem. But there are many unsolved problems in the life sciences. The surprise, to many in the scientific community, was the way in which he linked these to evolution itself, or at the very least, to the conception of nature held by biologists as part of the mechanism of evolutionary change. According to Nagel, if "materialist" science cannot explain consciousness, then evolution itself is very much in doubt.

News about the supposed "problems" of evolutionary theory is the routine stuff of science denial in the United States and other countries where the teaching of evolution is challenged for political or religious reasons. But the challenge posed by Nagel and some others, notably physician Raymond Tallis, author of *Aping Mankind*, is of another sort. Both Nagel and Tallis suggest that explanations of the evolutionary process based in the physics and chemistry of matter are not up to the task of accounting for the presence of human consciousness. And if evolution cannot account for consciousness, something else must be at work. But what?

THE NOT-SO-EASY PROBLEM

David Chalmers, whose image of a movie in our heads I employed at the beginning of this chapter, is well known for dividing the question of consciousness into two parts—the "easy problem" and the "hard problem."[4] The so-called easy problems of consciousness are the ones most directly connected to sensory experiences and our reactions to them. They include the ways in which we react to stimuli such as light, sound, touch, and smell. Other easy problems include the ways in which we focus our attention in time and space, the preconscious processing of visual images and sounds, the deliberate control of our own behavior, and the contrast between sleep and wakefulness. Chalmers calls these "easy" because the scientific pathways to solving them seem clear. Find the neural pathways that are activated when these processes take place, correlate them with awareness of the outside world, and you're on your way to a set of mechanistic solutions for conscious awareness. Chalmers is quick to point out that none of the "easy" problems is really that easy. In many cases, he speculates, they may take a century or more to solve. But none of these solutions, in his view, will require anything beyond the elucidation of the detailed neural mechanisms involved in conscious perception and awareness.

A set of experiments carried out by Stanislas Dehaene and his colleagues at the Cognitive Neuroimaging Unit in Saclay, France, illustrate one of the easy problems.[5] They first asked volunteers to report on what they saw when a pair of images were flashed on a screen before them. The first image was that of a digit, like the number 6. It appeared for just 16 milliseconds, then vanished, and was followed by a second image, which they called a "mask," showing four letters in nearly the same screen position as the digit. The image of the mask was presented for a much longer time, 250 milliseconds. The viewers were then asked to report what they had seen. Specifically, they were asked if they could recall seeing a number and, if so, which one.

The time delay between the first and second images proved crucial in determining whether the viewers became consciously aware of the number. If the delay was less than 50 milliseconds, most were unaware of the number and claimed they had not seen the first image at all. However, if the delay between the first image with the number and the masking image with the letters was more than 50 milliseconds, most of the viewers could accurately report which number they had been shown, indicating that by that time the information in that image had indeed risen to the level of their conscious perception. Clearly, a time delay had occurred between visual input of the first image and conscious awareness. If a second image is presented too quickly after the first, the first image does not make its way to the conscious mind. Instead, it is displaced by the second, and viewers are not aware of it.

The experimenters then sought to take these interesting observations a bit deeper by analyzing brain activity as the experiments were taking place. They found that regardless of the time delay, the first images always registered in a part of the brain toward the back of the head known as the visual cortex. Specifically, they recorded an *event-related potential*, or ERP, in this region of the brain after the first image was shown. If time delays between the first image and the mask were less than 50 milliseconds, the activity seemed to remain in the visual cortex and did not spread to other areas of the brain. However, once the delay between the first image and the mask rose to the point where subjects were able to report correctly on the number in the first image, neural activity became more widespread. Specifically, high levels of electrical activity were recorded in the prefrontal areas of the brain associated with conscious perception and analysis (these included the frontal, parietal, and occipital lobes).

Dehaene and his coworkers called this a *global ignition pattern*, in which large numbers of neurons produce action potentials, producing the ERPs recorded during their experiments. Summarizing these results, Dehaene wrote, "Clearly, conscious perception involves a massive amplification of the trickle of activity that initially arises from a brief flash of light. An avalanche of processing stages culminates at the point where many brain areas fire in a synchronized manner, signaling that conscious perception has occurred."[6]

These results are similar to those obtained in other laboratories that have examined what Christof Koch and the late Francis Crick called the *neural correlates of consciousness* (NCCs). They suggest that consciousness emerges from the synchronized firing of scores of neurons in the cerebral cortex. When sensory input rises to a certain point, it produces an "avalanche," to use Dehaene's metaphor, of electrical activity that spreads throughout the cortex and integrates multiple aspects of perception into a single conscious awareness. This, they suggest, is how we can pick up an object like a coffee cup and merge completely separate neural inputs regarding size, color, weight, and texture from different parts of the brain into a single conscious perception of that object so that we recognize it as a coffee cup.

Stepping back a bit, it is important to note that despite the sophistication of experiments like Dehaene's, the tools employed to analyze brain activity are still far too crude to pinpoint these NCCs in any meaningful sense. The changes in brain activity monitored in these experiments amount to electrical oscillations involving thousands or even millions of cells. Recording them as waves of activity effectively merges the very different actions of individual cells into a single whole. Far from tracing every link between a flashed image and conscious awareness, these experiments produce no more than vague impressions of which portions of the brain are involved in certain stages of the overall process. What is needed, as researchers know all too well, are precise tools that can report on the activities of individual cells and relate them, one step at a time, to the detailed activity of each element of the system. This is why the "easy" problem is not easy at all. But as more sophisticated neuroscience tools and techniques are deployed almost daily to attack the problem, we will come ever closer to an understanding of the physical nature of conscious perception. Researchers have every reason to believe they are on the right track, and more important, they have every reason to believe that the answers will come in scientific terms that are familiar to us—atoms, molecules, chemical reactions, membranes, cells, and electrical potentials. In fact, I'd bet on it. The remaining question is whether there is something more to consciousness these tools will never be able to probe. That's the really hard problem.

THE HARD PROBLEM

What is it like to be a bat? That's the question asked by philosopher Thomas Nagel in a classic 1974 essay of the same name.[7] In it, Nagel described the "hard problem" of consciousness decades before the term was coined by David Chalmers. Noting that conscious experience is widespread and occurs at many levels of animal life, he wrote, "an organism has conscious mental states if and only if there is something that it is to *be* that organism—something it is like *for* the organism." This, if you will, is *your* movie, the subjective nature of your own conscious experience. To sharpen his point, Nagel chose as his prime example a species close enough to humans that we can be sure it has conscious experience, but so different in its perception of the world that knowing what it is like to *be* that organism would be almost impossible. The bat.

Nagel describes bats as an essentially "alien" form of life, their perception of the world being entirely different from ours. Where we are visual creatures, bats are auditory ones, navigating swiftly, even in total darkness, by means of echolocation. Emitting a series of high-pitched chirps and shrieks, they detect the size, shape, location, even texture of objects by a form of sonar, using sound waves that bounce back to their ears. Their brains have an auditory cortex, as do ours, but theirs have regions in the cortex uniquely devoted to the spectral and temporal processing of sound. This means they devote a much larger portion of their brains to analyzing millisecond time delays from echoes bouncing off objects. They can also discern the shifts in frequency that indicate that an object is moving closer or receding into the distance. Their sonar, Nagel argues, is different from any sense we possess, and "there is no reason to suppose that it is subjectively like anything we can experience or imagine."[8]

Many years ago, when I read this essay for the first time, I wanted to quibble a bit. Humans can indeed echolocate objects, even though we're not very good at it. As a demonstration I've sometimes done for my students, I can blindfold a person in a large, quiet room and spin them around on a stool only a few feet away from a portable blackboard. When their spinning comes to an end, if asked to point in the direction of the blackboard, nearly every student does so accurately, despite the blindfold. The reason is that the sound coming off the spinning stool echoes back off the blackboard quite differently than it does from the open side of the room, and our auditory systems are good enough to detect that difference.

So, maybe in a limited way, we can imagine just a bit of what it is like to be a bat. But Nagel's point still stands, since the act of *being* a bat involves much more than an imprecise sense of echolocation. To truly understand, we would have to transform ourselves just as completely as Gregor Samsa, the protagonist of Kafka's *The Metamorphosis*, who awoke one morning to discover he had become a gigantic insect. We cannot, of course, and Nagel observes that "such an understanding may be permanently denied to us by the limits of our nature." Point taken, despite Franz Kafka's best attempts to slip the minds of his readers inside the exoskeleton of a bug.

SOLVING THE HARD PROBLEM

In a certain sense, Nagel's bat essay addresses one of the very real limitations of objective science. Scientists strive to take the personal point of view out of the process and analyze the world objectively. This is the very foundation of the scientific enterprise and has worked brilliantly for nearly every natural phenomenon. But this may not be the case for consciousness because of the truly subjective aspect of individual experience. For example, a person blind since birth has, of course, no understanding of the sensation of sight. Despite that, such a person can study electromagnetic radiation, learn which wavelengths are visible and which are not, and accurately describe light's interactions with matter. In short, a complete scientific understanding of the nature of light is available to someone who is blind. Nonetheless, the actual sensation of seeing white, red, or green light is not. No matter how precise the words, numbers, and formulae, science cannot replicate the sensation of sight so that a blind person might understand *exactly* what a sighted person experiences. An even more telling example might be to consider ultraviolet light, which we cannot perceive but many other creatures can, including some birds. What color do *they* see when they visualize an object in the ultraviolet?[9]

The obvious example of blindness invites a counterexample that might illustrate one way in which some have approached the hard problem. If you close your eyes in a darkened room and then place pressure on one of your eyes with a tap or gentle push, you may see an apparent flash of light. If this doesn't work for you, please don't hurt yourself trying to replicate the sensation, but I assure you it is as "real" as any sensation can be. It's a phenomenon common enough that James Dickey made it the centerpiece of one of his most remarkable poems, "Eye-Beaters." In it, a stranger visits a home for children and observes that some of the blind children have had their arms tied down so they can no longer strike their own eyes. Their longing to feel the sensation of sight in this way led some of them to beat their eyes repeatedly almost to the point of drawing blood. As Dickey wrote, "They *know* they should see." And then, "a long-haired nine-year-old clubs / Her eye, imploding with vision dark bright again again again / A beast, before her arms are tied."[10]

Dickey's imagination fashions something deep and primal about the urge to see, traveling backward in time to imagine an ancient and disturbing human past. Visions rise to the surface in the minds of these young eye-beaters as they struggle to understand the experiences their conditions have denied them. The visitor is so moved by the encounter that he must construct an understanding of these children in a way that changes him forever as he ends his visit and leaves the home.

The sensation of light produced by placing pressure on the eyes tells us something important about the nature of conscious experience. Sensations are physical events. Although not a single photon had hit a photoreceptor cell in the eyes of these children, the sensation of light was produced nonetheless. Sudden pressure on the eyes triggers a neural response that results in an actual sensation of light. This tells us that the sensation of light is indeed physical, and can be produced by physical means, even ones that have nothing to do with actual, visible light.

Consider what this might mean for the supposedly subjective nature of experience. It should eventually be possible to determine the precise neural circuitry that is activated when red, blue, or green light enters the eye and strikes photoreceptor cells in the retina. If we found a way, experimentally, to trigger the very same cells, we should be able to produce the actual sensation of those colors for a blind person. Such a procedure would, in effect, activate the neural coordinates of consciousness with respect to color perception. A blind person might then actually be able to see the difference between red and blue, experiencing the actual sensations we associate with such colors. If part of the hard problem lies in replicating the subjective nature of experience, surely this would be a provocative first step toward its solution.

Speaking of neural correlates, one of the most remarkable successes in the exploration of brain circuitry has come in the area of spatial recognition. It takes a while, we know, to get our bearings when we encounter a completely new space. Think of the first time you set foot on a large college campus or a complex multilevel shopping mall, or the first time you walked around a strange city. At first, you don't quite grasp the relative locations of different stores or buildings, but after a while you gain a sense of place that allows you to navigate. You come to understand where you are from a quick glance around and easily walk from one place to the next by the most direct route. How does this happen? How do we become familiar enough with a complex space to recognize our location within it?

In 2014, three scientists won the Nobel Prize for groundbreaking work that revealed how the brains of other animals pull this off. One of the prizewinners, John O'Keefe, at University College London, had discovered decades earlier a set of cells in the rat brain that become active whenever the animal enters a familiar space. Using electrodes implanted into a region of the brain known as the hippocampus, O'Keefe showed that these "place cells" were each associated with a specific location in the real world. When the animal found itself in a particular place, there was a specific place cell in the brain that fired, as if to tell the rest of the brain, "Hey—we are right here!" Pretty impressive, but it still leaves an important question unanswered. Place cells indicate recognition, but they don't explain how that animal constructed a mental map of relative locations that might have helped it move around from one spot to another.

The answer to that question was supplied by Edvard and May-Britt

Moser, two Norwegian scientists, who would go on to share the Nobel Prize with O'Keefe. Their worked identified a new group of neurons within the brain which they dubbed *grid cells*. As an animal becomes familiar with its location, these grid cells, laid out in a nearly hexagonal lattice, develop a two-dimensional pattern of neural firing that changes by moving along the neural grid as the animal moves in space. In effect, the grid cells contain a virtual map of that space, and their firing moves from grid cell to grid cell almost as if each cell were a single coordinate on a two-dimensional map.[11] It was a stunning discovery and provided a strikingly specific example of how cells within the brain become organized to generate a particular conscious experience—the sense of place.

Place and grid cells in the brain show us how the brain organizes its own cells to form neural representations of the world around us. As the tools of neuroscience improve, more discoveries will surely follow, chipping away at the hard problem by unraveling the ways in which the brain constructs perceptions and experiences. Gradually, its seems that more and more of the NCCs will become known, and the physical nature of the conscious experience will eventually yield to neuroscience.

Along those very lines, a team at the University of California, Berkeley, recently produced a "semantic map" of the human brain. Using a tool known as *functional magnetic resonance imaging (fMRI)* that detects neural activity by imaging increased blood flow, this team recently produced a "semantic map" of the human brain[12] by determining which portions of the cerebral cortex were activated when human listeners heard particular words in a story. They showed not only that certain words activated specific regions of the cortex, but also that these words activated the same regions repeatedly in a variety of human volunteers. Some areas responded to words about numbers, others to social words, and others to terms referring to places, shapes, colors, and so forth. A low-resolution tool, fMRI cannot reveal the activity of single cells, follow nerve impulses, or reveal precise connections between different regions of the brain. But it does demonstrate a path toward discovering locations within the brain that might be responsible for one of the most sophisticated aspects of human consciousness—the processing of language.

As a result of these and other studies, can we say with confidence that the hard problem will be solved by neuroscience? In a very important way, that depends on what one would accept as a solution. If finding the NCCs, the neural pathways, connections, and specific responses associated with conscious awareness would be a sufficient solution, then my answer would be yes. I can imagine, far into the future, an enormous, highly detailed map demonstrating each of the impulses and electrical potentials that are set in motion when I look at my kitchen counter and recognize a coffee cup. I can even imagine, somewhat further into the future, a much larger map of the circuitry that is activated when I look at that cup and decide that it needs to be washed out before I pour fresh coffee into it. Both are genuinely hard problems by any standard, and both can be addressed, I would argue, when the neural correlates of consciousness are sufficiently mapped.[13]

However, for a few neuroscientists and many philosophers, these advances, even if they eventually come to pass, would not be true solutions to the hard problem. Neuroscience cannot address the subjective, highly personal nature of the experience of seeing the cup, and certainly would not explain the judgment one person might make on whether the cup needs to be scrubbed out. Mapping the circuitry of consciousness, even in great detail, would not explain what it feels like to walk into the kitchen, see the cup, and come to the realization that it needs to be cleaned before use. NCCs might be able to explain how the sense of need for a cup of coffee is generated, but they will not be able to explain what that need feels like any more than they will be able to explain what it is like to be a bat. If we define the hard problem this way, then I suspect its solution is beyond the reach of neuroscience.

A PROBLEM FOR EVOLUTION?

In his book *Mind and Cosmos*, Thomas Nagel argues that an understanding of consciousness will not be found in evolutionary theory, or, as he puts it, in the "neo-Darwinian view" of the natural world. When I first read Nagel's book, this surprised me a bit. Surely consciousness is a problem for neuroscience, but why pick on evolution?

The reason, according to this respected philosopher, is that evolutionary theory expects to explain everything on the basis of physical law. Evolution, he writes, is based upon a purely reductionist viewpoint in which the physical sciences must be sufficient to explain the whole of nature. And since consciousness seems to lie beyond physical explanation, evolution does indeed have a problem. He puts it this way:

> Consciousness is the most conspicuous obstacle to a comprehensive naturalism that relies only on the resources of physical science. The existence of consciousness seems to imply that the physical description of the universe, in spite of its richness and explanatory power, is only part of the truth, and that the natural order is far less austere than it would be if physics and chemistry accounted for everything.[14]

We've already seen Nagel's well-reasoned argument that the subjective experience of individual consciousness lies, for the present at least, beyond the reach of science. I'll concede that much, but I am puzzled by his insistence that the natural order would be "austere" if it were based only on physics and chemistry. Perhaps he has a more dismal view of these disciplines than I do. But physics and chemistry fully account for the majesty of a sunset, the dazzling colors of a rainbow, the intricacy of snowflakes, as well as the sudden and terrifying beauty of a solar eclipse. Indeed, the gorgeous allure of the planet on which we live, so apparent when seen from space, is a product of repeating cycles of carbon, nitrogen, oxygen, and water. If this is austerity, give me more of it!

Despite this quarrel, Nagel is not alone in his insistence that consciousness presents a problem for the material view of nature. Physician, philosopher, and writer Raymond Tallis takes a similar point of view in his book *Aping Mankind*,[15] published two years before *Mind and Cosmos*. As Tallis sees it, neuroscience presents two big problems for any evolutionary explanation of consciousness. The first is the lack of explanation of how mental activity, including consciousness, could be based in matter. As he puts it, "The truth is, no theory of matter will explain why some material entities (e.g., human beings) are conscious and others are not." Since evolution "has to begin with matter and somehow end up with mind," it has "a problem with consciousness of *any* sort."[16] This is similar to Nagel's claim that the "neo-Darwinian" view of nature cannot account for consciousness.

Tallis's second objection echoes Nagel's doubts about the evolution-ary process itself. Nagel, giving voice to what many have called the "ar-gument from personal incredulity," asks, "what is the likelihood that, as a result of physical accident, a sequence of viable genetic mutations should have occurred that was sufficient to permit natural selection to produce the organisms that actually exist?"[17] As biologist H. Allen Orr pointed out in a review[18] of Nagel's book, "there's not much of an argu-ment here." But while Nagel's doubts, as Orr described them, rest only "on the strength of his intuition," Tallis has a more specific objection. He argues that the early stages in the gradual evolution of complex con-sciousness would not have been adaptive, which means they would not have produced an advantage natural selection could have acted upon.

I don't think Tallis has much of an argument here, either. He falls into the trap of demanding that every aspect of human intelligence, in-cluding "creating art or writing books," be "directly or indirectly related to survival, now, or at some time in the past."[19] As we've seen, this was Alfred Russel Wallace's concern as well, and it is effectively answered by the concept of evolutionary spandrels. Whatever forces caused the human brain to undergo its dramatic increase in size and complexity, the result was a mental instrument capable of much more than mere survival. We do not need to search for an ancient survival advantage that, directly or indirectly, accounts for our ability to do linear algebra, paint the *Mona Lisa*, or send rockets to the moon. Once evolution pro-duced the brains we now possess, they became capable of these feats and many more still to be imagined.

Tallis voices a further objection to the evolution of consciousness that strikes me as even more specious. He wonders "why evolution should have thrown up species with a disabling requirement to do things deliberately and make judgments."[20] He regards consciousness, especially in its early stages of development, as a burden that would have slowed down an individual organism's ability to react to life's chal-lenges. Referring to behaviors that might help to avoid predators, cope with environmental extremes, and compete with other individuals, he argues that nonconscious, "more tuned mechanisms" would be far bet-ter. What he overlooks is that while finely tuned, automatic-response mechanisms might be more useful in a certain set of highly specific situations, such behaviors would clearly lack the flexibility that goes along with consciousness. In an evolutionary sense, preprogrammed behaviors are brittle and less able to adapt to new situations and changing circumstances. The conscious and deliberative behaviors exhibited by many animals, including ourselves, not only make it easier to adapt to changing circumstances, but also better help to manage the prodigious amount of sensory information collected from the environment. Think of how much information is processed as you shift the gears of a car while navigating through heavy traffic, moving lane to lane, avoiding other vehicles, and reacting to unexpected situations. We certainly did not evolve to drive cars. But the flexible nature of conscious, deliberative behavior has enabled nearly all of us to be capable of acquiring that skill.

What seems to be going on here is that both authors, eager to claim that human consciousness is much more than a raw survival mechanism, fall into the trap of hyper-adaptationism. That is to say that each and every trait possessed by a living organism must have been favored by natural selection. So, if evolution would *not* have selected for consciousness, then consciousness must have been produced by some non-evolutionary process. Evolution thus fails to account fully for human nature.

This is the logic by which both Nagel and Tallis seek to preserve a sense of the uniqueness of the human animal and the idea that human thought and reason transcend their evolutionary origins. In particular, both are eager to defend the integrity of the human mind against the glib explanations of thought and behavior spun out by the excessive theorizing so common in evolutionary psychology. It's a well-intentioned effort, but it's not necessary.

Still, both authors acknowledge that degrees of consciousness are found in our closest animal relatives, one of the hallmarks of any trait with an evolutionary history. Tallis, for example, writes, " . . . higher primates are close to the threshold at which episodic self-consciousness passes over into sustained self-consciousness."[21] That, of course, is exactly what one should expect if self-consciousness were indeed an evolved characteristic. We see the beginnings of consciousness in other animals and its most complete development in ourselves.

Significantly, neither of these critics of an evolutionary explanation for mind or consciousness has an alternative theory. One might, therefore, be tempted to ask, if not evolution, then what? Since neither is an experimental scientist, both would surely assert that theorizing is not their job. Instead, their role is to stand outside of neuroscience in this regard and point to its flaws and misconceptions. Fair enough. But I sense that their real concerns are not with the evolutionary process itself or even with scientific materialism. Rather, they fear that the theories of neuroscience and evolutionary psychology diminish the human mind to a mere collection of atoms and molecules in a meaningless world of matter, energy, and change. Yet, as both have pointed out, such a conclusion actually undermines its own reliability by dismissing the scientific process that produced it. There must be, therefore, another way of approaching this, and so there is. It does not, however, involve rejecting our evolutionary nature, but understanding it more completely.

MATTER TO MIND

Earlier in this chapter, I alluded to the confusion that surrounds attempts to explain consciousness or even to agree on what *consciousness* means. That uncertainty persists, and there is no reason to pretend otherwise. So, in the absence of a complete explanation for the consciousness we all experience, how should we proceed? One starting point might be to agree that consciousness is based in the brain.

Like just about everyone my age, over the years, I've lost parts of my body and in some cases whole organs. I no longer have tonsils, an appendix, or a large portion of my right knee, but my consciousness persists, unimpaired. The same would be true, of course, if I should lose a kidney, a lung, or a limb. But lesion, injury, or death of the brain would be something else. Without the brain, and indeed without certain parts of the brain, consciousness is lost. Consciousness is not *identical* to the brain any more than sight is identical to the eyes, but consciousness is surely *dependent* upon the brain[22] in the sense that consciousness is one of the brain's principal functions. The brain is where consciousness lives.

Given this, what are we to make of assertions that the explanation of consciousness must be found in something other than the laws of physics and chemistry? Raymond Tallis writes that if you regard the brain as " . . . the seat of consciousness, then you are going to have to grant this bit of matter properties that no other material object possesses."[23] Tallis and Nagel both write as though they expect that consciousness requires something very different from the properties we ordinarily ascribe to matter. Maybe so. But if they are serious about this argument, it is only fair to ask where physics and chemistry fail. Is there a spot in the brain or elsewhere in the body where something is happening that is contrary to the laws of physics? Are there places where electrons change their charges from negative to positive, where ions flow against rather than with a concentration gradient, where energy is not conserved or the laws of thermodynamics are violated?

There are, of course, no such places, which is why so many in the field of neuroscience have been unimpressed by assertions that the phenomenon of consciousness violates our current scientific understanding of the natural world.

Surprisingly, even someone as critical of the materialist approach as Nagel sometimes makes an interesting concession to the physical reality of conscious events:

> So far as we can tell, our mental lives, including our subjective experiences, and those of other creatures are strongly connected with and probably strictly dependent on physical events in our brains and on the physical interaction of our bodies with the rest of the physical world.[24]

I'm not sure I could think of any more straightforward language to assert the dependence of mind upon body than these words of Thomas Nagel. And yet there remains something more to be considered. However thoroughly we convince ourselves of the physical nature of thought, we are left with a genuine problem. How is it that some collections, some groupings of matter are capable of mental activity and even consciousness while others are not? That is the philosophical problem that motivated these critics of evolutionary neuroscience in the first place. But it is a scientific problem, too, and one that speaks to the very core of the human experience as we would seek to understand it.

One way to approach the problem might be by thinking about the nature of life itself. A living cell engages constantly in commerce with the environment around it. Atoms are taken in, others pushed out, and all the while, powered by energy from food or sunlight, the cell builds new molecules while breaking down old ones. A plant inhales carbon dioxide and uses the energy from sunlight to "fix" those carbon atoms by linking them together to form carbohydrates. An animal inhales oxygen, using it to "oxidize" the very same sorts of carbohydrates, and exhales carbon atoms in the form of carbon dioxide.[25] These carbohydrates, like other compounds including proteins, lipids, and nucleic acids, are the principal components of living cells. When we consider a living cell, therefore, we attribute its properties to the collective actions of these and other molecules contained within its cell membrane. The study of these molecules and their actions is the business of one of the most important branches of biology, biochemistry.

All of this leads to a fundamental question, a bit more profound than it may seem at first glance. Is a carbon atom ever alive? Life is surely based in biochemistry, and a carbon atom that enters a living cell and then finds itself in a DNA molecule or a light-sensing protein is, at the very least, part of the substance of a living cell. Does this mean that something changes fundamentally in a nonliving carbon atom when it enters a cell and becomes part of that substance? As far as the most powerful tools of physics and chemistry can probe, the answer is no. The characteristics of atoms and molecules do not change when they become parts of a living cell, and the chemical processes that take place in such cells follow the same principles as those taking place in the nonliving world outside that cell. Nonetheless, there is something distinctly different about the matter in a living organism that is apparent to anyone who contemplates the difference between a rock and a blade of grass. Where does that difference come from? Can we take the fundamental principles of physics and from them predict how atoms will behave when organized into more and more complex groupings?

Physicist Brian Pippard thought this was not possible with respect to consciousness: "What is surely impossible is that a theoretical physicist, given unlimited computing power, should deduce from the laws of physics that a certain complex structure is aware of its own existence."[26] Like Raymond Tallis, who quotes this passage approvingly, I agree. But I would go a bit further, stating that it would also be impossible for that theoretical physicist to deduce that complex structures would take the form and properties of a living cell. By this I do not mean, nor did Pippard, that either cells or consciousness violate the laws of physics. Rather, that there are higher-level patterns of organization that, while based in those laws, are not completely explained by them.

If the opposite were true, if the properties of higher levels of atomic and chemical organization could be entirely derived from physics, we would have no need for a discipline like organic chemistry. When an industrial chemist synthesized a new compound, for example, she would have no need to determine its properties in terms of color, melting point, reactivity, optical activity, and so forth, because these could be precisely derived from those basic laws. In fact, chemists often have no better than a rough idea of what those properties might be, and such physical measurements are a necessary part of the analysis of any new chemical compound. The same is true for the much larger and more complex molecules studied by biochemistry and molecular biology. Complete upward determination from the laws of physics, however useful it might be, is simply not possible, and hence we must approach these more complex arrangements of matter top down rather than bottom up. In a sense, this is why biology is different from physics.

Life, therefore, is a phenomenon distinct from the physics of its own components. In even the simplest cell, the unique properties of life emerge from the collective actions and interactions of tens of thousands of different molecules. No wonder physics alone is not enough to describe it! No atom, by itself, is ever alive. But when atoms interact with innumerable others inside a living cell, those actions generate the remarkable process we call life. The same, I will argue, is true at an even higher level for the far more remarkable process of consciousness.

CONSCIOUSNESS EVOLVES

Philosopher Colin McGinn, despairing of a scientific explanation for consciousness, writes, "The more we look at the brain, the less it looks like a device for creating consciousness."[27] But just what should a device for creating consciousness look like? I'm not sure if McGinn would expect such a thing to have a little internal movie theater, a library of rules and past experiences, a rotating wheel marked off with decision-making possibilities or what. Since we know of no actual "devices" for creating consciousness, why would we expect it to have any particular form or structure at all?

Instead, I'd suggest we consider how living organisms are actually put together. Complex organisms, like ourselves, are built from cells. As many as 70 trillion of them in a typical human being. Some of these cells are solitary, like the cartilage-producing cells that dwell in the middle of *lacunae* (literally, "caves") surrounded by their own secretions. Others, like the epithelial cells that line the intestine, interact side to side with their nearest neighbors to form tight barriers that seal off compartments in the body. Some, like muscle cells, are filled with contractile fibers and are bundled into work gangs (the muscles themselves) awaiting the signal to contract and produce movement, swift and sure. In each case, the structure and organization of cells gives clues as to their function. So, what would a cellular organ for creating consciousness look like? A hyperconnected, hyperactive bundle of 86 billion cells with high-speed sensory inputs and high-speed motor outputs, perhaps? If so, that description certainly fits the human brain.

The greater question, of course, is how consciousness arises from those cells. Neuroscience may be able to identify the actual cells associated with consciousness, the NCCs, but can we explain how those cells differ from other neurons we do not consider conscious? Or what special property enables some forms of matter to become conscious while others are not? The first step in approaching these questions is to realize that consciousness itself is not a "property" of either molecules or cells, any more than life is a property of the carbon atoms in that candy bar I just ate. Matter itself does not *become* alive. Rather, certain groupings of matter are capable of generating the tangled complexity of a self-sustaining process we call life. Consciousness, similarly, is not a property of matter or even a property of individual cells. In a way analogous to life itself, consciousness is a process generated by the hugely complex interactions of highly active cells within the brain and associated nervous tissue. Consciousness, therefore, is something that matter *does*, not something that matter *is*.

This insight doesn't explain human consciousness any more than the realization we are made of the same chemical stuff as the rest of existence explains life. But it does tell us where to look. It tells us to keep working, to keep teasing one secret after another out of the overwhelming jungle of connections that generate conscious awareness. And it tells us that if we ever approach an answer, it will be based, just like life itself, in the wonders that ordinary matter works within the confines of the living cell.

Why do some neurons (and their associated cells) seem to generate consciousness while others do not? Why, if all action potentials (nerve impulses) look pretty much the same, do some generate perceptions of sound, others of smell, and others of vision? Why are some brains capable of conscious self-awareness while others are not? One might as well ask why all data streams in the region of the electromagnetic spectrum used by cellular phones look nearly identical, though some of them carry spoken conversations, others carry text, others carry pictures, and still others are coded instructions that can reprogram the very devices receiving those signals. How does such diversity arise from a single stream of electromagnetic radiation? The answer, of course, lies in the structure of the receiving device that detects, decodes, and represents each class of data differently. Engineers and programmers can explain these details for cellular data streams. Before too long, neuroscientists may be able to do much of the same for neural activity.

Raymond Tallis objects to this as a possibility, tagging such expectations with the label "neuromania." By *neuromania*, he means "the appeal to the brain, as revealed through the latest science, to explain our behavior."[28] One of his examples is vision. He notes that while the experience of seeing the color yellow is triggered by radiation in a certain region of the spectrum (with wavelengths around 570 nm), the sensation of "yellow" is distinct from the radiation itself. As a result, he argues, one cannot account for the sensation of seeing yellow in terms of neuroscience, since it involves the very same sorts of ion flows and electrical changes that are triggered by other stimuli, such as heat, cold, touch, and the color blue. Therefore, no investigation, no matter how thorough, of the neural pathways involved in seeing the color yellow can fully explain how the experience is different from the seeing the color red or smelling a rose.

In terms of the subjective inner experience, point taken. But objections like Tallis's overlook of the fact that nervous systems react to stimuli by creating representations of those stimuli in the only ways available to living cells. In the brain, these are nerve impulses, rapid changes in ion flows, and more long-lasting changes in cellular connections. Think, for example, of the grid cells that become organized into a multidimensional map, a neural representation of a complex space in the "real" world. It's no revelation that a tiny cluster of grid cells is not the same as the real-world space it represents. But the discovery of these cells showed how the brain constructs a functional cellular representation of such a space and explains how neurons produce a sense of place that helps us navigate through the real world. Similar work will surely show how the brain represents "yellow" or "blue" or a high-pitched sound. I may not be able to know if your inner sensation of yellow is the same as mine, but we can surely investigate the chemical and cellular processes that produce those inner sensations.

Recent work has shown how the brain handles a perception far more complicated than color—facial recognition. Primates, ourselves included, are especially good at distinguishing individual faces, and previous research has shown that this ability appears to be centered in a region of the brain known as the inferotemporal cortex. Two researchers at Caltech, Le Chang and Doris Tsao, monitored the firing of individual "face cells" in this region as macaques were shown a series of computer-generated human faces. Their data showed that as few as two hundred of these face cells were required to identify a particular human face.[29] Furthermore, individual cells responded to specific combinations of facial dimensions, which the researchers were able to determine by varying specific features in each face, and then determining which cells changed their firing patterns. Eventually the researchers reached a point at which they could predict the neural firing patterns that would result from individual faces. They were able, in essence, to read the neural code for faces. Commenting on her work, Dr. Tsao noted that many researchers had begun to think that the brain might be a "black box" in which it might be impossible to learn the actual mechanisms of perception. "Our paper provides a counterexample," she stated. "We're recording from neurons at the highest stage of the visual system and can see there's no black box. My bet is that that will be true throughout the brain."[30] I would make the same bet. The activities of the brain, including perception and consciousness, are based in the workings of its cellular components, and explicable in purely scientific terms.

What then of the two great problems that consciousness is said to pose for evolution?

The first is that consciousness requires an explanation beyond our current ("neo-Darwinian") understanding of nature, which is to say, beyond the accepted physics and chemistry of today. While we have only begun to lift the veil of mystery that envelops the workings of the human brain, it is also clear that we have found nothing within it that contradicts or violates our understanding of natural processes. In many respects, Thomas Nagel's argument that current science is not up to the challenge of explaining consciousness echoes arguments from many decades ago regarding genetics. In the early 1940s, the great physicist Erwin Schrödinger addressed the problem of biological information in a series of lectures, later published under the title *What Is Life?* [31] Doubting that biology was up to the task of explaining genes, the basic units of inheritance, Schrödinger wrote that a fundamental advance in our understanding of nature was required to explain the molecular nature of genes. Specifically, Schrödinger predicted that science had to open itself to the possibility that "other laws of physics, hitherto unknown," would be required to solve the problem.

To be fair, unlike today's critics of neurobiology, who find science fundamentally flawed in its reliance on physical principles, Schrödinger also wrote that these "other laws," of physics, "once they have been revealed, will form just as integral a part of this science as the former."[32] And Schrödinger's prediction? A generation of scientists who came of age in the post–World War II era, including James D. Watson of double helix fame, were inspired by Schrödinger's provocation to identify and characterize DNA, now widely understood as the molecule of the gene. There are, it is safe to say, no "new" laws of physics resident in the DNA molecule. Instead, there is in it an unexpected level of structure and function that enables DNA to serve multiple functions of coding and copying in ways that pre–double helix physics and chemistry could not have anticipated or predicted. Physics itself remained intact after the discovery of the double helix. What changed was our appreciation of the potential of the physics and chemistry of matter. The same is surely true of the physics and chemistry of thought and consciousness.

The second supposed problem for evolution concerns the possibility that the evolutionary process might generate beings like us—and, for that matter, like bats and dolphins and orangutans—all capable of subjective conscious experience. This strikes me as a particularly weak objection. As Nick Lane writes in his book *Life Ascending: The Ten Great Inventions of Evolution*, "If the mind is not a product of evolution, what actually is it?"[33] Is there some other, mysterious force that produced the conscious mind? We've already reviewed the many lines of evidence documenting the evolutionary origin of the human species. The same sorts of evidence apply to most of the other species we might regard as conscious, and authors such as Tallis and Nagel do not challenge that evidence. It is also clear, as we saw in earlier pages, that the brain underwent a dramatic increase in size and complexity during the emergence of our species, and this may have forced an extensive rewiring of critical connections within the cortex related to conscious self-awareness.

The real difficulty that Tallis, Nagel, and others have with the idea that consciousness is a product of evolution has little to do with the actual evolutionary process, to which they present no alternative. Rather, they are uncomfortable with a certain sort of evolutionary thinking that, in their view, degrades and dehumanizes the species. Nagel makes this explicit. He is a "moral realist" and does not see how a utilitarian, survival-related "Darwinian" explanation of the human moral sense could be compatible with that understanding. As he puts it, "since moral realism is true, a Darwinian account of the motives underlying moral judgment must be false, in spite of the scientific consensus in its favor."[34] The alternative Nagel does not consider is that evolution may have produced a mind capable of finding what he regards as authentic and objective moral values in the same way that it produced a mind that became capable of finding truth in science, mathematics, and even in art. In short, he falls into the adaptationist trap of concluding that everything produced by evolution exists only to serve the logic of survival and nothing more. I would suggest the entire history of our species shows otherwise.

Tallis does not invoke moral realism *per se*, but in the opening pages of *Aping Mankind*, he suggests that neuroscience and the theory of evolution, "two of our greatest intellectual achievements," have been used "to prop up a picture of humanity that is not only wrong but degrading."[35] Apparently provoked by another writer who rejects the idea that there might be anything distinct, different, or noble about the human animal in his critically acclaimed book (*Straw Dogs*, by John Gray), Tallis, like Nagel, regards a "Darwinian" explanation of human consciousness (as well as a physical or cellular one) as threatening and dehumanizing. Motivated by such concerns, he trivializes any attempt to equate brain activity, those flows of ions and electrical charges, with consciousness. The work of the conscious mind, he argues, is something that defies such purely physical explanations. He writes that he cannot deny what is staring him in the face, "namely that we are different from other animals and that we are not just pieces of matter."[36]

On these points, Tallis is absolutely correct. We are not *just* pieces of matter, and we certainly are different from other animals. But in haste to make these critical points, he devalues both the process (evolution) that has produced those differences and the physical basis (neurobiology) of why we are much more than *mere* pieces of matter. He notes, as I have, that all those arguing against a special status for the human animal are, in fact, humans themselves. Other creatures do not make such arguments and even lack the capacity to pose the question in the first place. That alone makes us special.

What of the complaint that brain science can offer nothing more than a mundane accounting of neural connections, ion flows, and action potentials with no real significance? To argue that the base language of membrane potential and ion flows cannot account for consciousness is like saying that the mere stringing together of twenty-five or thirty symbols in a linear sequence cannot account for the drama or poetry of Shakespeare, or that certain two-dimensional arrangements of pigment on canvas cannot stir the soul in the deepest sense, because so many other arrangements amount to mere decoration or pattern. The notion that complexity emerges from simplicity is in fact a recurring theme in nature and in life. Sound rises to music, words to literature, and cells to organisms. Neural impulses by themselves are surely like single notes, individual letters, or solitary cells. By themselves, they cannot capture or represent the depths of reality. But taken together, they construct a tapestry as rich with nuance and meaning as the world itself.

The certainty, expressed by many in the philosophical community, that conscious thought and experience must be fundamentally beyond the reach of material explanations offered by science is clearly misplaced. This is not because they overstate the wonders of consciousness, but because they *understate* the capabilities of the physical world. Philosopher Galen Strawson makes exactly that point when he writes that many make the mistake " . . . of thinking that we know enough about the nature of physical stuff to know that conscious experience cannot be physical. We don't." He continues, " . . . the hard problem is not what consciousness is, it's what matter is—what the physical is."[37]

Ironically for those who think consciousness could not have evolved, the link between coded systems of neural impulses and connections to an authentic reality has actually been forged by evolution itself. As Nick Lane writes, "Feelings feel real because they have real meaning, meaning that has been acquired in the crucible of selection, meaning that comes from real life, real death. Feelings are in reality a neural code, yet a code that is vibrant, rich in meaning acquired over millions or even billions of generations."[38]

We need not worry that our status as creatures of evolution degrades the reality of conscious experience, reduces human thought and logic to the meaningless movements of atoms and molecules, or renders pointless the great achievements of human art and culture. And we certainly do not need the claim of mystery to find genuine humanity in the workings of the human mind. Evolution has shaped the conscious self, to be sure, but it has shaped it in a most remarkable way. It has given us perceptions so refined that we have become aware not only of our surroundings, but of ourselves as well. It has also given us the motivation to consciously seek to understand the life within us and the tools to do so.

Chapter 7

I, Robot

C offee or tea? Should I open a book after dinner or watch TV? Will I take the highway or the back roads on my way to work today? Not a day goes by, in fact, not a conscious minute goes by, when we don't make choices. Most, like these, are mundane and of little consequence. But some are profound. We vote in elections, we choose whom to marry, whether to have children, and whether to obey the orders of a policeman. We can reason, come to conclusions, and form opinions on art, music, and politics. In one way or another, we are defined by the choices we make, the values that guide our thinking, and the ways in which these choices affect our lives and the lives of others.

But what if you were told you had never made a single authentic choice in your entire life? What if you came to realize that every word you uttered, every step you took, every thought you believed you had was actually determined by forces beyond your control? And if you did come to realize that, what would you do about it? Would you try to break free? Or would you come to understand that even the very thought that you were being programmed was beyond your control, and therefore even that conclusion was suspect? That is the great question posed by the issue of free will and, in the opinion of many, by evolution as well.

In his book *The Moral Animal*, Robert Wright declares that "free will is an illusion, brought to us by evolution." He goes on to say that those things we are either blamed or praised for "are the result not of choices made by some immaterial 'I' but of physical necessity."[1] That "necessity" reflects the combined results of an evolutionary process that endowed us with powerful instincts and a nervous system governed by physics and chemistry rather than the power of individual choice. As biologist David Barash put it, "Thus, to my mind (and I believe I write this of my own free will!), there can be no such thing as free will for the committed scientist, in his or her professional life."[2]

Barash and Wright are hardly alone in claiming that accepting the reality of evolution implies an end to the concept of free will. As the late Cornell philosopher Will Provine said, in a talk celebrating the work of Charles Darwin, "Naturalistic evolution has clear consequences that Charles Darwin understood perfectly," including that "human free will is nonexistent."[3] Cambridge philosopher Stephen Cave, noting "The sciences have grown steadily bolder in their claim that all human behavior can be explained through the clockwork laws of cause and effect," traced this "shift in perception" to the "intellectual revolution" ushered in by Darwin's work, *On the Origin of Species*.[4] Cave also points out that after its publication, Darwin's cousin, Francis Galton, interpreted evolution to mean, "our ability to choose our fate is not free, but depends upon our biological inheritance."[5]

Darwin himself never addressed the issue of free will directly in any of his published writings, although a 1838 notebook the twenty-nine-year-old Charles did refer to free will as a "general delusion."[6] Much later, in *The Descent of Man*, Darwin considered the clash between our more primitive evolutionary instincts and the nobler imperatives of social convention: "At the moment of action, man will no doubt be apt to follow the stronger impulse; and though this may occasionally prompt him to the noblest deeds, it will more commonly lead him to gratify his own desires at the expense of other men."[7] While this is a fair description of the conflicts we all feel from time to time between what we want to do and what we feel we should or must do, it is hardly a rejection of free will. Nonetheless, evolution deniers routinely bring up the issue of free will, and many are surely persuaded that the loss of freedom is just one of many grim consequences wrought by the Darwinian worldview. Philosopher Daniel Dennett, a tireless promoter of evolutionary explanations for human nature, conceded this point in his book, *Freedom Evolves*: "Concern about free will is the driving force behind most of the resistance to materialism generally and neo-Darwinism in particular."[8] Are such concerns valid? Or is there a way to explain free will within an evolutionary framework? Dennett thinks so, and so do I.

AN EASY CASE

The question of whether our actions are genuinely free is hardly a new question. Such debates go back at least to the Stoic school of philosophy in ancient Greece and perhaps much further. The issue has remained contentious for so many centuries, in fact, that one of our contemporary philosophers has described it as a "scandal" that so little progress has been made in settling the question.[9] If you are seeking a definitive solution to the problem, therefore, it might be wise to look elsewhere. Nonetheless, we can still ask whether our evolutionary origins imply an absence of free will and whether the workings of our highly evolved nervous systems make free will a possibility.

While we all feel as though we have free will—presumably, you chose to read the words on this page—we also have to consider one of the basic principles of science, which is that all events have causes. The tides rise and fall. Why? Must be a cause somewhere. Rainbows appear just after a storm passes. Milkmaids almost never get smallpox. Lung cancer is more common in smokers than in nonsmokers. How come? The search for cause, the hunger for explanation is the very heart of science, and therefore we seek causes for everything human as well.

If human thoughts and actions are to be explicable by science, they must have causes, including ones encompassing the very decisions we instinctively believe to be freely made. It therefore becomes easy, almost too easy, to make a case against free will. As Samuel Johnson wrote, "All theory is against the freedom of the will; all experience for it."[10] We feel that we are free from the moment we decide to rise until we turn in at night, and we experience that freedom intimately, directly, personally. As John Locke wrote, "I cannot have a clearer perception of any thing than that I am free."[11] Indeed. But might that perception itself have a cause? And, if it does, how can the sense of freedom be taken as anything but illusion? For that reason, if no other, a case against free will is far easier to make than a case for it.

I realized this many years ago in high school, when our English teacher asked members of the class to choose sides for a debate on free will. While nearly all my classmates picked the affirmative, I grimly chose the negative. It was the 1960s, individualism was very much in flower, and my debate opponent quickly invoked that spirit to make his case. "We are all individuals," he explained, "and as such, each of us is different, each of us unique, and each able to think for themselves. That wouldn't be possible without free will." These were fine sentiments, I remember thinking, but they lent themselves to a quick rebuttal.

Our classroom was in a new wing of the school that had been finished just that year. I gestured toward the newly painted cinder-block walls. "Every one of those blocks is different from every other block. But does that mean they have free will? Of course not. Being different doesn't imply freedom. If it did, every prisoner in the state penitentiary would be just as free as you and me, because they are all different, too." My classmate continued to struggle in vain to support free will, and as he did, all I had to do was to point out the failings of one argument after another. I argued that every action has to have a cause, and if our own actions do have such causes, then our will cannot possibly be free. Hardly an uplifting conclusion, but it was an easy case to make, and it won the day.

THAT GHOST, AGAIN

The reality of free will seemed especially clear to René Descartes, the great seventeenth-century mathematician and philosopher. He wrote that freedom of the will is "self-evident," and "must be counted among the first and most common notions that are innate in us."[12] However, Descartes also saw the body and even the brain in mechanistic terms. He repeatedly described the body in such language, referring to it as "a machine made of earth"[13] and writing that it was "not necessary to conceive of this machine as having any vegetative or sensitive soul or other principle of movement and life."[14] However, Descartes also believed that no mere machine could possess free will, which he regarded as an essential human characteristic. To deal with the apparent contradiction between free will and the mechanistic nature of the body, Descartes needed a place in which the rational self might be grounded. For that purpose he chose the pineal gland, a tiny structure near the center of the skull at the base of the brain. Why the pineal? Part of the reason may have been that he couldn't think of any other function for this tiny organ, which today we understand as part of the endocrine system.[15] Descartes himself, however, put forward a quite specific anatomical rationale for the special role he assigned to the pineal. Remarkably, he based his assessment on a quirk of anatomical symmetry:

> My view is that this gland is the principal seat of the soul, and the place in which all our thoughts are formed. The reason I believe this is that I cannot find any part of the brain, except this, which is not double. Since we see only one thing with two eyes, and hear only one voice with two ears, and in short have never more than one thought at a time, it must necessarily be the case that the impressions which enter by the two eyes or by the two ears, and so on, unite with each other in some part of the body before being considered by the soul. Now it is impossible to find any such place in the whole head except this gland.[16]

Seeking a place in the body to house the freedom of will he found self-evident, Descartes was clearly taken with the pineal's apparent unity of structure, so different from the bilateral organs of sensation. Since our thoughts are unitary, the unitary pineal must be the locus of free will, which Descartes identified as essential to the soul. Having defined the immaterial mind and the material body as two completely different substances, he also needed to identify a point of interaction, so that the mind could be aware of the body's sensations and so that the body could respond to the will of the mind. This form of dualism sets the mind—and the soul—apart from the mechanistic rules governing the body and solves the problem of free will by placing it beyond the capacity of material science to investigate.

Nearly all philosophers, and certainly all neuroscientists, have long since discarded Cartesian dualism. Nonetheless, the critics and debunkers of free will still seem to find it necessary to attack any notion that a nonmaterial mind might be resident somewhere within the human body. For example, E. O. Wilson writes, "The brain and its satellite glands have now been probed to the point where no particular site remains that can reasonably be supposed to harbor an nonphysical mind."[17] Curiously, Wilson seems to miss the point that if such a region could be found within the brain, it would be physical in every sense. Nonetheless, he is eager to let us know that Descartes was wrong, and neither the pineal gland nor any other region of the brain is home to that ghost in the machine, the nonphysical mind. Steven Pinker is even more explicit when he asks, "How does the spook interact with living matter? How does an ethereal nothing respond to flashes, pokes, and beeps and get arms and legs to move?"[18] How, indeed?

Pinker's by now unremarkable point is that the brain is physical, not ethereal, and that we need not concern ourselves with anything other than the physical workings of the human brain in our search for the sources of human will and decision making. As he put it in a widely viewed online video:

> I don't believe there's such a thing as free will in the sense of a ghost in the machine, a spirit or a soul that somehow reads the TV screen of the senses and pushes buttons and pulls the levers of behavior. There's no sense that we can make of that. Our behavior is the product of physical processes in the brain.[19]

Arguments like those of Wilson and Pinker seem more directed toward shielding science from spiritualism or a ghostly vitalism than addressing the issue of free will itself. Bound up with their concerns is the question of soul, which they associate with free will and which they identify with religious superstition and even bigotry. Pinker believes we have the ability to "select" certain actions by predicting their consequences, implying that this may "carve out the realm of behavior we call free will," but he dismisses any suggestion this might involve "some mysterious soul."[20] Sam Harris's critique of free will as resident in the soul is more direct:

"Few concepts have offered greater scope for human cruelty than the idea of an immortal soul that stands independent of all material influences, ranging from genes to economic systems."[21]

In nearly every respect, modern critiques of free will embrace a determinist view of nature. That is, that all events, including those within the human nervous system, are the results of previous events and conditions. Therefore, they are, at least in some sense if not totally, predetermined. Determinism is closely linked to the idea of cause and effect, without which science as we know it would be impossible. Therefore, to many, accepting any notion of free will would be akin to rejecting science itself.

If science requires that we exclude magic and mystery from the realm of human thought and choice, then determinism seems to be the only choice. We don't want to throw up our hands and run away from the question, leaving that ghost to haunt the machine. As a result, we can view the case *against* free will as essentially a case *for* science, pure and unadulterated. By this logic, to qualify as part of science, evolution seems to require the denial of free will.

WHY DID I WRITE THIS?

While the case against free will seems easy to make, full acceptance of a determinist view of human behavior leads to some very odd situations. Many years ago, I attended a lecture by a noted evolutionary biologist in which he made it clear that the concept of free will amounted to a soul-based spiritualism that had no place in science. Given recent advances in neuroscience, that lecture might have seemed even more compelling today. But in the question-and-answer session, he was asked about the greatest challenges facing the planet. Without hesitation, he said "the preservation of biodiversity." What followed was a brief, passionate sermon on the need to retain wild spaces, conserve the habitats of endangered species, and pass as much as we could of the great diversity of life along to future generations. Naturally, I agreed with these conservationist sentiments, as would any biologist.

Then he pointed out to this audience of scientific specialists that it was incumbent on all of us to use our training, our positions, and our professional skills to build a social and political consensus around this issue. Great advice. But here was a person who had just rejected free will on scientific grounds asking all of us to make a conscious choice to persuade others, who presumably also lack free will, to accept a scientific consensus on the preservation of biodiversity and change their actions accordingly. Presumably, that speaker felt not only that he was free to express that opinion, but also that he had arrived at it on the basis of careful, reasoned, deliberative decision making informed by scientific findings. And yet his previous remarks had proclaimed, with equal passion, that all of this was determined by conditions and forces over which neither he nor his listeners actually had any control.

While many discussions on free will focus on immediate practical issues such as the social systems that reward achievement or punish criminal behavior, there is a deeper issue that transcends these considerations. If thought and behavior are strictly determined by a set of internal and external conditions, then even our own views on free will itself are predetermined. How, then, should we approach this question and even the more general question of how we decide what is true? As I mentioned at the beginning of this chapter, this is not a new problem. Lucretius described it well more than two thousand years ago:

> If all movement is interconnected, the new arising from the old in a determinate order—if the atoms never swerve so as to originate some new movement that will snap the bonds of fate, the everlasting sequence of cause and effect—what is the source of the free will possessed by living things throughout the earth?[22]

For this reason, most philosophers do not consider free will an easy question and have developed a wide range of views on the subject. Some of these involve adjusting the definition of free will as described a bit earlier by Steven Pinker. Others fall into the "compatibilist" school, holding that free will is compatible with determinism, and our decision-making abilities allow us to make rational choices between alternatives. Neuroscientist Stanislas Dehaene describes his view of free will in such terms:

Our brain states are clearly not uncaused and do not escape the laws of physics—nothing does. But our decisions are genuinely free whenever they are based on a conscious deliberation that proceeds autonomously without any impediment, weighing the pros and cons before committing to a course of action. . . . What counts is autonomous decision making.[23]

Nonetheless, Dehaene's "autonomous" decision making isn't exactly "free." To him, those "voluntary" decisions are "ultimately caused by our genes, our life history, and the value functions they have inscribed in our neuronal circuits."[24] His descriptions walk the compatibilist tightrope between strict determination, finding both external and internal causes for our actions, but still allowing for the conscious deliberation we seem to sense every time we make up our minds.

By contrast, Sam Harris, the author of bestselling books such as *The End of Faith* and *Letter to a Christian Nation*, is frankly uninterested in finding even a degree of compatibility between scientific determinism and free will. He writes, "Free will is an illusion. Our wills are simply not of our own making. Thoughts and intentions emerge from background causes of which we are unaware and over which we exert no conscious control. We do not have the freedom we think we have."[25]

Harris cites experiments[26] indicating that brain activity associated with decision making takes place well before we become conscious of making a choice. Therefore, even those actions we think to be purely voluntary, like raising our hand in class or picking which of a dozen eggs to use in a recipe are actually predetermined by nonconscious processes within the brain. For Harris, the absence of free will is good news. It will lead to a more enlightened view of how to mete out punishment and structure rehabilitation for criminals once "their culpability begins to disappear." It will reform politics by doing away with the myth of the self-made man and allowing each of us to see the conditions that brought us to our various stations in life. For Harris, the ultimate value of discarding free will is that "Getting behind our conscious thoughts and feelings can allow us to steer a more intelligent course through our lives." But wait a moment. Harris pauses, apparently aware that he is about to contradict himself by citing the virtues of "deciding" that you do not have the freedom to decide anything. He then attempts a rescue in the form of the parenthetical afterthought that one can choose to steer that intelligent course, "while knowing, of course, that we are ultimately being steered."[27] The logical contortion here is striking.

Even more remarkable is the way in which Harris concludes his brief, pamphlet-length book on free will. Throughout the manuscript, he has been at pains to explain that he does not know why he makes certain decisions. These include why he just drank a glass of water as opposed to a glass of juice. "The thought never occurred to me."[28] Why coffee this morning as opposed to tea? "I am in no position to know."[29] Why, after a hiatus of more than twenty years, did Harris again take up martial arts training? He cites a book he read on responding to violence. But he has no answer as to why he found that book compelling or why he now practices two martial arts, confessing, "the actual explanation for my behavior is hidden from me."[30] Harris no doubt thinks these small concessions to mystery do not undermine the case he is striving, with considerable eloquence and skill, to make against free will. But one great question looms if one does indeed accept his arguments. Why did he decide to write a book to convince his readers, who lack the free will to decide anything, that in fact they should conclude, as he has, that free will is an illusion?

A genuine answer to this question would be interesting, but Harris can do no more than throw up his hands and walk away. Toying with his readers, in the final three pages of his narrative, he tells them he will write "anything I want for the rest of the book."[31] That includes putting a "rabbit" or an "elephant" into one of his sentences. How will he decide which of these two words is better? He has no answer. Is he free to change his mind about those words? Of course not. Because, as he puts it, his mind can only change him. So, why does he end his book at this point? Well, he tells us he is hungry as he writes that final page. That is a feeling he can resist, of course, but only for a while, and in any event, it is time for the book to conclude. Once again, why? No answer other than the "feeling" he has made his point, and, as Harris walks away to find something to eat, one is left to ponder whether he has indeed made his case. But even that conclusion, as Harris would be sure to remind us, is not really ours to make.

No matter how one makes the case for biological determinism or lays out the argument against free will, there remains an uneasy logical problem behind that line of reasoning. That is, of course, that if we lack free will, then scientific logic itself is no longer valid. We cannot claim to make decisions or draw conclusions on the basis of evidence, we cannot pretend that scientific investigation is a path to truth, and we cannot even justify writing a book to get people to "believe" in the absence of free will. The reason is that belief in anything is not a free choice, but an artifice of genetics, circumstance, and uncontrollable external stimuli. This was painfully apparent to physicist Stephen Hawking as he penned the opening pages of one of his popular books on space, time, and the cosmos:

> The ideas about scientific theories outlined above assume we are rational beings who are free to observe the universe as we want and to draw logical deductions from what we see. In such a scheme it is reasonable to suppose that we might progress ever closer toward the laws that govern the universe. Yet if there really is a complete unified theory, it would also presumably determine our actions. And so the theory itself would determine the outcome of our search for it! And why should it determine that we come to the right conclusions from the evidence? Might it not equally well determine that we draw the wrong conclusion? Or no conclusion at all?[32]

Acceptance of behavioral determinism undermines not only itself, but all of science and perhaps the arts and humanities as well. It is stunning how few critics of free will seem to realize this and to appreciate the grim nihilism that flows from such ideas.

BROKEN CLOCKWORK

If the denial of free will brings problems, then surely an assertion of free will brings problems of its own. Critics of free will might rightly demand to be shown how a nervous system—or any natural system, living or mechanical—can do something that is not based on a definite cause or collections of causes. Find a mechanism that can do something not dictated by its preexisting state. In other words, find an uncaused cause. You're looking for a spook, the critics might say, because there is no way out of cause-and-effect determinism.

Before retiring in despair, however, let's consider what physics actually tells us about the nature and behavior of matter. As the nineteenth century was drawing to a close, there was some confidence that a fully deterministic program for physics was nearing completion. The laws of motion were well understood, electricity and magnetism had been united into a single electromagnetic force, the atom had been dissected into its basic components, and even the nature of light seemed ready to yield its secrets. Indeed, at the beginning of that century Pierre-Simon Laplace had articulated a philosophical position that now seemed reasonable. That was that if a single intellect could, at the present moment, know the position, mass, charge, and velocity of every particle in the universe, it could predict the exact state of the universe at any point in the future or determine its state at any point in the past as well. As Laplace put it, "for such an intellect nothing would be uncertain and the future just like the past would be present before its eyes."[33] In such a world, the absence of free will would be an unassailable fact of science.

But as the twentieth century began, it became clear that we do not live in such a world. Relativity and the discovery of quantum effects led to a revolution that upended the strict determinism of Newtonian physics. We now know that the behavior of matter at its most basic level, the elementary particle, is profoundly nondeterministic. While the average behaviors of large numbers of particles such as electrons can be predicted quite well in a statistical sense, the movements of individual particles cannot. This means that strict determinism does not hold and that Laplace's elegant idea of divining both past and future from time present falls apart. Putting it more bluntly, this means that every detail of our present universe was not determined from the moment of its origin. Right turn on red, disco music, and the designated hitter rule were not inevitable consequences of the big bang.

Free will would be clearly impossible in a deterministic universe.

We are creatures of matter and energy, and subject to the laws of physics. If the universe played out like a windup machine, so would we, and our behavior would be just one more part of that predetermined clockwork. But it does not work that way, and neither do we. We share the indeterminate nature of the quantum world, and therefore can be subject to its effects. For some, this serves as a way to rescue free will from the trap of determinism. At the very least, it means that we are not subject to the kind of strict predictability that would have made free will impossible. Rather, our contemporary understanding of physics leaves that possibility open.

Can we find a way to wire indeterminacy into the circuitry of the brain in a way that affects choice and behavior? At a simplistic level, yes. We could begin by noting that the brain is much more than an input-output device that integrates sensory signals and sends out hardwired command impulses in response. In fact, the vast majority of neuronal activity within the brain is internal, self-generated, and independent of direct sensory input. I can, for example, close my eyes and recall the sight, sizzle, and scent of a rack of ribs roasting over a charcoal fire, no actual sights or smells required. I can choose to imagine a parabolic curve plotted on a Cartesian coordinate system, and carefully show to myself that the slope at any point of the curve $y = x^2$ is equal to $2x$, one of the most basic operations in elementary calculus. These are thoughts I consciously choose to have, and they are based in neuronal activity independent of the outside world.

Just as significant are those activities taking place without my conscious awareness, and these dominate the moment-by-moment workings of billions of brain cells, surely including some of those in decision-making pathways. Given that the all-or-none firing of any particular neuron is triggered by ionic currents that can be tripped one way or another by the unpredictable movements of electrons and soluble ions, the actual behavior of even a relatively simple neural circuit becomes unpredictable, even in principle. Therefore, the notion that complete knowledge of the location of every cell, every protein, and every ion would allow behavior to be strictly determined is every bit as incorrect for the brain as it is for the universe as a whole. If a mere escape from strict determinism is all you ask of free will, we already have it.

This is not, however, a very satisfying formulation of free will. It attributes our choices and decisions to the tosses of molecular dice embedded deep within consciousness, and not subject to anything resembling truly voluntary action. If this is all there is to free will, one might wonder whether it was worth having, let alone worth arguing about. There has to be something more, and perhaps there is.

MORE IS DIFFERENT

In 1972, physicist Philip W. Anderson wrote an article for *Science* magazine entitled "More Is Different."[34] In it, he addressed the conception, popular in the science of that time, that all of science is explicable in terms of the most basic properties of matter and energy. Define these fully and one could apply a kind of "constructionism" to explain ever more complex systems including molecules, structures, living cells, and even whole organisms. Anderson, a leading figure in physical science who first described what has come to be known as the Higgs mechanism to account for the mass of elementary particles, was awarded the Nobel Prize for Physics in 1977. In his article, Anderson soundly rejected the notion that science could work upward from a few basic laws to describe and predict the behavior at higher levels of organization accurately. Rather, he pointed out that the constructivist project breaks down when confronted "with the twin difficulties of scale and complexity." Instead, new approaches are needed at higher levels of organization:

> The behavior of large and complex aggregates of elementary particles, it turns out, is not to be understood in terms of a simple extrapolation of the properties of just a few particles. Instead, at each level of complexity entirely new properties appear, and the understanding of the new behaviors requires research which I think is as fundamental in its nature as any other.[35]

Anderson's point, for which he provided specific examples, was that at each stage of organization, "entirely new laws, concepts, and generalizations are necessary." "Psychology is not applied biology, nor is biology applied chemistry." I made a similar point in chapter 6, citing emergent properties that attend the greater levels of chemical and biological complexity associated with consciousness. We might therefore ask whether the indeterminate nature of quantum-level events might make it possible for something like free will to emerge as a higher-level property of a complex nervous system.

ENGINES OF CHOICE

We aren't the only organisms that make choices. Critters as simple as bacteria make them all the time, and they can be matters of life and death. One of the critical choices many bacteria make is in which direction to swim. Naturally, a bacterium swimming around in a liquid would like to move toward a source of food rather than away from it, and most do this quite effectively. The technical term for this is *chemotaxis*, since cells move toward food or other *chemoattractants* in their growth medium. How do they do this? How do they choose the "right" direction that will lead them toward ever-increasing concentrations of such chemoattractants? You might say that they harness the power of randomness.

Many bacteria swim by means of flagella, little whiplike structures they twirl in unison to pull or push them through liquid. An individual bacterium swims by twirling its flagella in a counterclockwise direction. This causes the flagella (between four and eight in a typical cell) to bind together into a rotating bundle that pushes the cell powerfully along. Every now and then, however, the flagella stop and reverse the direction of their spinning. This causes the bundle to come apart, and the cell tumbles and spins randomly. Just as suddenly, the flagella reverse again, twist into a single bundle, and the cell moves forward once again, this time in a completely different direction.

How does this help the cell swim to increasingly higher concentrations of food? As it turns out, it has a sensory system that monitors the concentration of chemoattractants. If the cell is swimming through higher and higher concentrations, it tumbles less often. If the concentration is getting lower, it tumbles more often. In this way, it adjusts the direction of its swimming time and time again to move toward attractants like food and away from chemical repellents. The mechanism combines a necessary element of randomness with a kind of choice to produce actions that benefit the organism. Could something like this be operating in the human brain? Is there a way to harness unpredictability to make conscious, sophisticated choices that are not strictly determined in advance?

A number of scientists have given this possibility serious consideration. One of the first to do so was the physicist Roger Penrose. In his 1989 book *The Emperor's New Mind*,[36] Penrose suggested that the brain itself might function according to quantum mechanics in a way that generated both consciousness and free will. His ideas were not well received by most physicists, some of whom noted that true quantum computing depends upon hardware cooled to within a few fractions of a degree of absolute zero. Under such circumstances, the quantum state of an individual particle can directly influence the output of a computing device. By contrast, the warm, wet, and noisy environment of the brain would not be an ideal place for quantum-level effects to have any noticeable influence on the workings of neurons. Some years later, however, Stuart Hameroff, a physician interested in the workings of the nervous system, took up Penrose's suggestions and proposed that certain structures within neurons might provide places where quantum phenomena could indeed influence cellular activities.

Hameroff points to observations indicating that the firing threshold for brain neurons is not quite as narrow as described by standard neurobiological theory and actually varies over a wide range. This means, he suggests, that something else, an x-factor, is actually determining whether or not a neuron will fire given a specific stimulus. He claims that other factor might well be conscious choice, mediated by a process known as *quantum entanglement*. In plain language, something happens in the brain that allows it to regulate its own activity. In an effort to pinpoint a mechanism for this to take place, Hameroff points to microtubules, supporting structures that make up part of the neuron's internal cytoskeleton and are also found in nearly all cells with nuclei. Microtubules, as the name implies, look a bit like hollow microscopic straws. They self-assemble from a protein known as tubulin and act almost like railroad tracks to guide the movement of materials from one end of a neuron to another.

Hameroff suggests that certain chemical groups within the tubulin protein can engage in a kind of quantum entanglement that spreads a wave function among microtubules in multiple neurons.[37] Quantum computations among these entangled neurons terminate in a process Roger Penrose called *objective reduction*, which triggers neuronal firing and control conscious behavior. These processes solve the problem of free will by "sending quantum information backward in classical time, enabling conscious control of behavior." This backward causality addresses the problem of how a neuronal activity can cause itself, since objective reduction would, in effect, allow events at the present time to reach backward, affecting the system's current state. By allowing actions at one instant in time to determine the state of the brain itself at the very same moment, a pathway to genuine free will could be charted.

As intriguing as this complex (and difficult to explain) proposal might be, critics have not been kind to it. Some have argued that the specific effects upon which it would depend occur too quickly or too rarely in biological systems to give credence to such speculation.[38] Others have pointed out that there is no experimental evidence for quantum entanglement in living cells and asked how it is that microtubules in neurons, but not in other cells, would have acquired such properties. Finally, the notion of an event becoming its own cause presents philosophical as well as scientific problems that Penrose and Hameroff have not been able to resolve.

A much less fanciful approach to the question has recently been developed by Dartmouth neuroscientist Peter Ulric Tse. This approach, described in his book *The Neural Basis of Free Will*,[39] is based on a highly detailed understanding of the plasticity and flexibility of the nervous system itself. There is a tendency, common among those who write about free will as philosophers or physicists, to treat the brain as hardware. It is often described as an immensely complex machine, intricately interconnected and driven by sensory inputs, whose mechanistic workings are strictly determined by action potentials passed from one neuron to the next. But Tse points out that the brain is actually dominated by synapses, and so "the state of a neural network might be better described by specifying the state of its synapses than the firing pattern of its neurons."[40] This matters because the strength and sensitivity of synapses on both sides of a connection between two neurons can be rapidly altered by neural activity itself. Therefore, the ability of neural activity to modify the "hardware" of the brain can actually change the way in which the brain acts.

The plasticity of the brain has been known for some time, of course. For example, a 2009 study showed that learning to juggle induced changes in the white matter of a region of the brain associated with vision and motor skills (the *intrapareital sulcus*).[41] Tse's own laboratory has also shown that learning Chinese induces similar changes in the language-processing circuitry.[42] You might say these and other studies confirm the wisdom of the Greek philosopher Heraclitus, who, as quoted by Plato, said, "You can never step into the same river twice." The river is changed by your presence, and so is the brain changed by everything it experiences, including one's own thoughts.

Tse cites research indicating that spikes of neural activity can change, within milliseconds, the sensitivities of the synapses through which they pass. This means that a single neuron could be unresponsive to a stimulus that, just a fraction of a second ago, was sufficient to trigger a distinct nerve impulse. This rapid reweighting of synapses could alter neural pathways just as decisively as throwing a couple of switches can alter the destination of a train by switching it from one track to the next. For this reason, neural systems don't operate merely by sending coded messages from cell to cell; they actively switch the routing of those messages so that neural systems can change the nature of their responses from one moment to the next. As a result, the brain itself sets the criteria for its own future activity on an almost instantaneous basis.

This is where Tse locates the mechanism of choice associated with free will. He agrees that at any instant, the way in which the brain will act is deterministic, specified by the detailed conditions of the neural networks and systems at that point in time. However, the activities of neurons at that very instant can and do change the nature of the network for future actions, including those that may take place within milliseconds. Now, this is not quite enough to provide for genuine free will since, as Tse points out, even a zombielike creature could use the mechanism of synapse reweighting to modify future actions. What is distinct about his proposal is that conscious activity itself can drive synaptic reweighting. Therefore, neural activity associated with conscious choice can change the state of the system, resulting in a different response in the future that remains deterministic at the moment, and therefore does not violate the basic principles of scientific cause and effect.

Tse proposes a "three-stage neuronal model" to account for free will and choice. In the first stage, rapid synaptic resetting of a neural network occurs in response to preceding mental processing. Next, variable inputs arrive and are processed according to criteria in the newly reset network. Third, the reset network fires or does not fire according to these criteria. Tse suggests that randomness can occur in the first two stages of this model, but not in the third. As a practical matter, this means that to some extent randomness does play a role in generating new patterns of neural firing that then govern the deterministic response that occurs in the third stage. Tse calls this phenomenon *criterial causation*.

Tse claims that criterial causation gets around one of the usual objections to free will, which is that no process can cause itself. The processes he has described surely do not cause themselves, but rather create the causes for future actions in ways that are subject to one's conscious will. His colorful (and classical) description is that criterial causation "offers a path towards a strong free will that passes between the unfree 'Scylla' of determinism and the equally unfree 'Charybdis' of randomness."[43] As he put it in a popular article explaining his ideas, "our brains can set criteria, play events out internally, choose the best option, then make things happen."[44] By most reasonable standards, that's very close to the sort of free will we inherently believe we possess. So, has Tse solved the problem?

Maybe. He surely has gone further than anyone else in his attempts to locate a genuine mechanism in neural systems that could be responsible for making choices according to certain criteria. Because mental events such as thoughts are also physical events at the level of the cell, he has pointed out ways in which thought itself can affect the future activity of the brain. This may satisfy certain definitions of free will, and it certainly provides a mechanism by which a choice made at one time can influence subsequent actions. To be sure, those initial choices are strongly constrained by genetics, experience, and the preexisting state of the nervous system, as Tse admits. Of course, our choices are influenced by these and other factors, and even a strong conception of free will does not require otherwise. Nonetheless, as I consider Tse's ideas, I wonder whether all he has done is to describe a far more elaborate, sophisticated, and adaptive sort of determinism. Given that immediate decision making in his system is still determined by prior conditions, that is a fair criticism. But what he has done should surely give pause to those who think the scientific case against free will is closed. It is not, and there are good reasons to keep thinking about it.

SECRET STRINGS?

Many writers, including Sam Harris,[45] confidently assert that laboratory experiments have now shown conclusively that free will is an illusion. They base these claims on a famous series of experiments initiated more than three decades ago by neurophysiologist Benjamin Libet. In a typical version of Libet's experiment, volunteers were asked to perform a simple act, such as tapping a finger or pushing a button, whenever they wished. They were also asked to watch the movement of a dot moving quickly around a clocklike device and note the position of the dot at the exact moment they felt the urge to act. Since his subjects were also connected to an electroencephalogram device, Libet was able to monitor brain activity leading up to each tap of the finger. His instrumentation detected a rise in brain activity in the motor cortex, part of the brain that controls physical motion, about 500 milliseconds *before* his subject reported the urge to act. Similar experiments have been carried out many times since, and most have observed such increases in brain activity, called *readiness potentials*, prior to conscious awareness of a decision. These observations have been interpreted to imply that even decisions as simple as when to move a finger are determined "for" us by nonconscious regions of the brain over which we have no control.

More recent experiments carried out in the laboratory of John-Dylan Haynes in Berlin indicated that the timing of such decisions might actually come as much as seven to ten seconds earlier,[46] and it might even predict whether the subject was going to push a button with their left or their right hand. As persuasive as these experimental results may seem, their implications for free will are questionable.

As Daniel Dennett has pointed out,[47] there is a little more to these experiments than it would seem. He asks us to consider that the brains of participants do not "see" and record the position of the moving dot as a single, instantaneous event. The pathway from the retina to the visual cortex takes several milliseconds, and it may take even a bit longer for visual processing to present the image to the conscious self. These delays then have to be added to the time it might take for the brain to make the decision to push Libet's button, and then to make a judgment as to which "snapshot" of that moving dot was truly simultaneous with the decision. But then the perception of a decision may be further delayed by the time required to activate the motor cortex to send instructions to motor neurons to trigger muscular contractions that actually cause the finger to move.

Finally, as Dennett explains it, as our brains send motor commands to our bodies to carry out actions like rapidly typing a word on a keyboard, the time delay between the decision to type out a particular letter and the motor actions of the finger in pressing the corresponding key is such that the brain has learned to monitor it with a built-in time delay. In other words, milliseconds elapse between the mental command to press the letter k and the actual time at which our fingers execute that command. In the meantime, if we are rapidly typing the word *keyboard* (as I am right now), more commands are quickly being issued for the letters that follow (*e*, *y*, *b*, etc.), and these go out before my fingers actually complete the task of pressing the letter k. Therefore, our brains have been trained to monitor such actions by comparing the tactile and visual feedback from typing not with current command decisions, but with those that were made several milliseconds ago. Nonetheless, we perceive the command and the pressing of the letter to be nearly simultaneous. Dennett suggests there is every reason to believe that the very same time delay is part of Libet's experiment, so the gap between readiness potential and conscious awareness of the decision "is an artifact of mis-imagined theory, not a discovery."[48] As he puts it, "You are not out of the loop; you *are* the loop."[49]

Dennett's criticisms of these experiments have been supported by recent experimental findings pointing out that the so-called readiness potential does not actually indicate a subconscious decision. Instead, these fluctuations of brain potential may indicate the beginning of the decision process or the necessary preparation of the cortex to initiate movement.[50] Peter Ulrich Tse and his collaborators have joined this emerging consensus with highly sensitive experiments showing the lack of any "causal relationship" between readiness potentials and the awareness of a willingness to move. While the readiness potential may be an interesting part of the decision-making process, it does not indicate that our conscious wills are no more than puppets on a string. The truth, as it often is, is surely more complicated—and much more interesting.

FREEDOM EMERGES

I have no doubts that debates on free will are going to continue for a very long time. They began with philosophy, migrated to physics, and now have spread to neurobiology. If anything has changed over this time, it may be that philosophers today cannot logically approach the subject without at least a passing acquaintance to the science that increasingly defines the limits of philosophical reasoning and speculation. I take that as a good thing, even though I have some confidence that before long a final answer will emerge from the rapid progress we are now seeing in the field of brain science. As interesting as it might be to hold back on any conclusions until we can actually identify the networks or cellular clusters where decisions are made, there is another question more directly related to the theme of this book that we can approach. That is, what does evolution tell us about free will?

As we have seen, evolution, like all science, is based on a view of the natural world that sees life as a physical process. If one insists that physical processes are bound by strict predictability, then the ability to make free choices is strictly an illusion. If, on the other hand, one notes the inherent unpredictability of such processes, then there is at least the possibility of free will built into living systems, even if we cannot be sure how or where it might be generated. What we can agree upon, I suggest, is that free will, if it is not real, is at the very least a powerful illusion that affects the ways in which we think and act. As biologist David Barash puts it, even those who reject free will (as he does), nonetheless "experience our subjective lives as though free will reigns supreme."[51] Whether illusion or reality, the sense of free will is powerful and compelling. So, where did it come from?

Daniel Dennett asks us to consider the nature of choice as it exists for us and for other creatures. If one defines freedom as the ability to choose between alternative paths of action, then we surely have infinitely more degrees of freedom than a bacterium spinning wildly to orient itself in a chemical gradient. A fruit fly likewise exceeds that bacterium in terms of the possible pathways of action, and a bird encounters this sense of freedom to a still greater degree. But as Dennett says, none of these can approach the possibilities for choices of action that attend a human being.

> We are the only creatures whose members can *imagine* the adaptive landscape of possibilities beyond the physical landscape, who can "see" across the valleys to other conceivable peaks. The mere fact that we're doing what we're doing—trying to figure out whether our ethical aspirations have any sound anchoring in the world science is uncovering for us—shows how different we are from all other species.[52]

Where did this capacity come from? How did we arise to see such possibilities, to judge the alternatives, to reason our way through the world? The answer is evolution. Psychologist Roy Baumeister echoes this view when he writes, "The evolution of free will began when living things began to make choices."[53] Human beings are distinguished from other animals by virtue of our highly developed, verbally based culture. We have the ability to articulate the reasons for our choices to others, enabling us to overcome evolutionary drives and instincts by conscious acts of the will. These are the basis for human social interactions, and they demonstrate that free will is much more than a mere illusion. It is the basis for our success and even our dominance of this planet, for better or for worse, as a species.

There is no need to invoke magic or spiritualism or to confer intentionality upon individual atoms to reach this understanding. Baumeister puts it this way:

> We cannot break the laws of physics, but we can act in ways that go far beyond physical causation. No electron understands the Golden Rule, and indeed an exhaustive study of any given atom will furnish no clue as to whether it is part of a person who is obeying or disobeying that rule.[54]

While the free-will sentiments of Dennett and Baumeister are not universally shared in either the philosophical or scientific community, it is striking to note what their critics believe to be the ultimate sources of the "illusion" of free will. Science writer Rita Carter spoke for many when she tied the human sense of freedom to our evolutionary past:

> The illusion of free will is deeply ingrained precisely because it prevents us from falling into a suicidally fatalistic state of mind —it is one of the brain's most powerful aids to survival.[55]

Carter added that the illusion of free will affects society by engendering a sense of personal responsibility that gives rise to both legal and social punishments for aberrant behavior. Whether one regards such strictures as positive or negative is a value judgment, of course, but there can be no doubt that the assumption of free will underlies our attitudes toward crime and punishment. Free will, therefore, provided an apparent sense of personal responsibility that served us well as human societies began to develop and thrive.

Interestingly, as Raymond Tallis has pointed out, the idea that free will is a mental illusion produced by evolutionary natural selection has an ironic twist. If we truly lack free will, then our conscious thoughts and desires are determined by forces beyond our control, and we cannot change future events by choice. Yet, if the "illusion" of free will did have adaptive value, then it actually did change the course of events by helping human social groups to cohere and prosper. Therefore, if free will is an illusion, it is an illusion that became self-fulfilling.

Perhaps Lucretius was right, and atoms sometimes do swerve to "break the bonds of fate." Perhaps, deep within us, there is a mechanism of the sort described by Roger Penrose or Peter Ulric Tse that can harvest the unpredictability inherent in the physical nature of matter to produce free choice and provide a scientific basis for free will. Perhaps, as Daniel Dennett has argued, we are fully deterministic creatures in which free will as a rational choice between possible alternatives derives from the complexity and self-consciousness of the human nervous system. Or, as the critics would have it, perhaps free will is an illusion, an evolutionary hangover that helped us to muddle through as a social species to become the dominant animals we are today. No matter to which school of thought you subscribe, you will find evolution right at the center of any explanation of free will, whether genuine or illusory. Darwin is not the enemy of free will. For, if we are indeed truly free, it was evolution that made us so.

Chapter 8

Center Stage

I 'm hoping for a clear sky tonight. It's expected be the peak of the annual Perseid meteor shower, a chance to glory in streaks of sudden fire as fragments of a comet come crashing through the Earth's atmosphere. In between those moments of spectacle, there will be a chance to lie still in the darkness and absorb the quiet beauty of the nighttime sky. The experience has always made me feel small against the vastness of space, but it's also one that has helped me, as a biologist, appreciate what it means to be human. Although I expect to be alone in my small backyard, I'm not the only one who will be looking up tonight. Tens of thousands of people will be watching around the world, a quiet and widely scattered assembly of those who remain fascinated by such events year after year.

Consider the setting. Joined by these many others, I'll lie back against the surface of a small, rocky planet, peering up in wonder at the twinkling riot of forms and colors and patterns. The sparkling fire of the meteors is new, generated only a fraction of a second before it flashes across the sky. The tapestry of starlight, however, is a sampling of history, some of it unimaginably ancient. I orient myself by Polaris, the north star, fully aware that the stream of steady light it provides is more than four hundred years old. Sirius is much brighter, owing to its nearby position. Its light took just eight and a half years to reach me.

But deep in the sections of sky that look fuzzy and indistinct to my eyes are whole galaxies whose light was sent on its journey billions of years ago. We are only left to wonder what has happened to them since? Do they still exist? Have they withered into darkness or exploded into even greater glory? The answer is on the way. We have to wait only a few billion years to find out.

Of all the creatures, of all the forms of life that grace the surface of this small planet, there is only one that looks this way into the nighttime sky. Only one knows the Perseid spectacular is coming. Only one plots the distances to stars. Only one contemplates the age of its universe, only one is aware of the mysteries to be solved in starlight. While all of life is one, while all of life is linked by ancestry, structure, and design, only the human creature seeks answers to questions in the stars. This is what makes it worthwhile to consider how this creature came to be, and what its presence on this planet means.

ADAM'S PROMISE

For people in Western cultures, the character of Adam once defined the essence of human nature both in promise and tragedy. As Marilynne Robinson has noted, the story of evolution brought on the collapse of the Genesis narrative, and with it, in her view, the enlightened humanism that produced Western civilization and gave birth to the very science that would, ironically, lay waste to the myth of Eden itself. To her and many others, Adam was much more than a pseudo-explanation for the origin of our species. He was the metaphorical source of man as a moral creature with obligations to family, community, and ultimately to the righteousness of truth. While evolution is surely true, as Robinson admits, what it put in Adam's place was hardly a satisfactory image to replace these fine qualities:

> For old Adam, that near-angel whose name means Earth, Darwinists have substituted a creature who shares essential attributes with whatever beast has recently been observed behaving shabbily in the state of nature. Genesis tries to describe human exceptionalism, and Darwinism tries to discount it.[1]

As I indicated much earlier, I think Robinson is fundamentally wrong about the implications of what she calls "Darwinism." That is, in fact, my reason for writing this book. But she is surely right about the conclusions many have drawn from the emerging story of human evolution. We could begin with the very exceptionalism she tries so valiantly to defend. Earlier, we saw how Henry Gee, in his book *The Accidental Species*, also discounted such exceptionalism. Nothing, he wrote, is uniquely present in our species, including attributes such as language, toolmaking, intelligence, mathematics, or even self-awareness. So, we have no reason to presume ourselves special, unique or, as Gee gleefully points out, the "pinnacle of Creation." We're just not that big a deal, and we have no business thinking otherwise.

Gee's gospel of insignificance blends seamlessly into Stephen Jay Gould's contingent view of history in which neither human beings nor mammals nor even vertebrates were the inevitable products of evolution. Rewind the tape of life, using Gould's metaphor, and something very different would have emerged. Evolution was not bound to produce us or anything like us. As a result, any notion that evolution results in progress is a self-gratifying illusion. The drama of evolution plays out not in an irresistible rise to perfection, but in a random walk through endless possibilities, none more significant than the other, none especially worthy of our attention.

Inherent in this view is the importance of randomness in the process. Evolution is driven by a tension between the unpredictability of genetic variation and the demands of natural selection, which is a manifestly nonrandom process. Gould emphasized the importance of historical accident as a decisive factor in driving the currents of natural history, and therefore his view of the rise of humanity is akin to a shrug. We are an historical accident, not the result of a grand design, nothing more than the children of luck and happenstance. To think of ourselves as special, therefore, is an error of judgment and logic.

If these constructions tend to devalue human life just a bit, in the eyes of many interpreters of "Darwinism," there are even more depressing findings to deal with. Our bodies, our minds, our behaviors have all been shaped by the harsh demands of survival in the face of the relentless pressures of natural selection. As a result, however sophisticated we may seem, we are rude creatures at heart, motivated by drives and values that serve principally to propagate our genes and ensure our own survival and that of our kin. As Richard Dawkins once wrote, "Let us try to teach generosity and altruism, because we are born selfish."[2] The endowments of evolution apparently include a surplus of ruthless greed and aggression, but a deep and telling absence of love and kindness, virtues that, according to Dawkins, are not inherent in our species and can only be passed along by deliberate effort.

In the 1980s, when I first became active in public efforts to defend evolution, I thought, quite naïvely, that debates with "scientific creationists" and their kind would be settled by factual arguments. In truth, their talking points about a 6,000-year-old Earth and gaps in the fossil record were laughable and easily refuted by showing the real evidence. What I did not expect was the tenacity with which evolution deniers would stick to their beliefs in the face of so much science to the contrary. Particularly striking to me, at first, was their insistence on a literal Adam and Eve, on the initial perfection of creation, on the sins of that first couple, and on a literal Fall from grace into the death and chaos of life as we know it today. I could not see why the historicity of the Genesis story, which even many ancient Christian scholars did not take literally, was held so closely and with such passion.

One answer, often presented to me, was that if the Bible could not be trusted to be true in its very first book, then how could its more recent books, the Gospels, be taken as the word of God? This was a response I tended to shrug off, noting that the Bible is not a single book but a collection, a mini-library of books written by various authors at different times for a variety of audiences and often for very different purposes. So, the veracity of one book chosen (by men) for that library said nothing about the accuracy of others written a thousand or many thousands of years later. But I sensed there was something even more meaningful in the Genesis story, something well beyond simple slogans embedded in lines like "Adam was my ancestor, and not some chimpanzee,"[3] which is part of a song that some creationist organizations teach to children. These deeper concerns transcend the doctrinal demands of biblical literalism and speak directly to the human self-image. Who are we? Where did we come from? And what are we capable of?

To all too many, the answers that emerge from the Darwinian narrative are dark, foreboding, and deeply unsettling. First among these is the conviction that our minds are not our own. They surely were not formed in the image and likeness of a supreme being, and they were not even fashioned in a way that allows us to seek the truth of our own existence. Rather, our brains are organs like any other, only one component of a survival machine designed to resist death just long enough to push its genes forward into the next generation of struggling, highly socialized primates. Evolutionary psychology can explain our moral values as instinctive behavioral patterns hewn only by selection for life within the group. Art is made to attract mates, altruism is practiced for selfish reasons, even if we "think" otherwise, and "truth" is a constructive illusion connected only loosely to an unknowable reality. Freedom of thought and action is part of that illusion, a lie the brain tells itself to allow the human animal to function in a way that enhances its chances of success. High culture is not the work of genius, but the product of chance adaptations working in many brains to sculpt a veneer of beauty around the mundane realities of life and struggle. Beauty itself is defined only by its ability to produce such illusions as allow us to go on under the absurd circumstances of personal futility and ultimate death.

Seen by those who would explain every impulse, from anger to joy to love, in Darwinian terms, the human project seems worthy of neither pride in past nor hope for the future. If even consciousness is an illusion, then it is pointless to contemplate that future, seek wisdom in the past, or celebrate human achievement. By contrast, the myth of Adam once affirmed a genuine humanity. It told us that choices were freely made, that their consequences were genuine, and that rebellion made possible by truly independent thought was an essential part of human nature. It was for such reasons that Marilynne Robinson lamented the "death of Adam" in terms like these:

> Our hypertrophic brain, that prodigal indulgence, that house of many mansions, with its stores, and competences, and all its deep terrors and very right pleasures, which was so long believed to be the essence of our lives, and a claim on another's sympathy and

courtesy and attention, is going the way of every part of collective life that was addressed to it—religion, art, dignity, graciousness.[4]

While certainly not a creationist in the sense of denying evolution, Robinson perfectly articulates the profound concerns of those who recoil from extremes of the "Darwinism" she describes in such chilling terms. But her view of evolution as a denial of human nature, as a nihilistic project that devalues not just religion, but art, music, literature, and even science, is, I believe, profoundly wrong. What evolution tells us about human nature projects an entirely different vision of our species. It invites us to revel in the living world of which we are a part and to see ourselves as central characters in the greatest drama the universe has yet brought forth. It is a story that fully matches the sense of grandeur with which Charles Darwin once tried to endow his greatest theory, and we should delight in telling it.

THE ASCENT

The creationist view of natural history has always struck me not only as wrong, but as deeply impoverished. Their image of an Earth formed instantaneously and then immediately populated with all the great diversity of life is static, inflexible, unchanging, and, frankly, boring. Our planet's genuine history is a long-term narrative, and an epic one at that. Even the continents are dynamic, not fixed but drifting as tenuous crusts above a deeper molten sea that shudders and shakes and occasionally breaks through to fire ash into the air and pour rivers of lava into the sea. Life has mastered this planet's perilous geology, slowly at first, but over time with increasing power and effect, changing the Earth and its atmosphere in dramatic ways that mark ours as a planet of life. If there are distant aliens whose instruments have gathered a telling glimmer of sunlight reflected from the sun's third planet, they already know that something exceptional has happened here. The highly reactive oxygen in our atmosphere, the infrared light reflected from forests and fields of living plants, the very temperature of the planet all reveal that life has taken over this small planet.

And what a takeover it has been. Once established, life evolved. It

diversified, it moved to new locations, it established new ways of extract-
ing energy and raw materials. In the process, one form of life branched
off from another and then branched yet again as the whole of life explored the enormous range of adaptive possibilities. The cell, the basic
unit of life, proved so adaptive that it could join with others to produce
larger organisms that would invade the land and produce whole new
ecosystems, now populated by organisms that could crawl and run and
even fly. Life accomplished all of this while retaining a basic unity at the
level of the cell that speaks to its common ancestry, its shared history,
and its universal reliance on a set of common chemical principles and
building blocks. All of life is truly one.

While creationists base their natural history on separating humankind from nature, evolution tells an entirely different story, one
of unity with the natural world. As a result, when Ernst Haeckel and
others depicted the historical relationships of different species, they
explicitly included humans as one of the branches on a large tree,
linked directly to the whole of life (see Haeckel's diagram in Figure
3-1). Today, of course, we look at Haeckel's formulation a bit more
critically, for two fundamental reasons. First, its title was *The Pedigree of Man*,[5] but given the nature of the diagram, a more fitting title
would be "the pedigree of all living things," since it purported to show
all major living groups within a single tree of life. More important,
biologists see no reason to place human beings at the very top of the
tree, as if we were the "highest" evolved form, standing at the very
pinnacle of life itself. A more telling (and accurate) diagram is shown
in Figure 3-2, where the multiple branches of life radiate out into a cir-
cular pattern, and there is no implication that any one form or type is
higher or lower than any other. As Darwin himself once wrote, "never
say higher or lower."[6]

In biological terms, this is surely the correct way to regard our
own relationship to a living world to which we are but one small twig
among many millions of branches, and a relatively recent twig at that.
As Henry Gee and many other have emphasized, there is no reason
to regard our own biological endowments as superior to those of other
creatures, many of whom have physical or biochemical qualities far superior to our own. We are just one of many.

But that analysis holds only if we think narrowly enough to confine our gaze to mere biology.

In a 1973 book and television series, the mathematician and historian of science Jacob Bronowski described *The Ascent of Man* in language that still resonates. Bronowski's intent was to describe the way in which our species has risen to dominance on the planet, and also to trace the intellectual pathways that have led to the development of human culture, art, civilization, and especially to science. In the very first chapter of his book, entitled "Lower Than the Angels," he began to lay down the case for human exceptionalism:

> Man is a singular creature. He has a set of gifts which make him unique among the animals: so that, unlike them, he is not a figure in the landscape—he is a shaper of the landscape. In body and in mind he is the explorer of nature, the ubiquitous animal, who did not find but has made his home in every continent.[7]

Bronowski's narrative was overarching in terms of time and space. It began in the great Rift Valley of eastern Africa, the home of our species. It highlighted the first truly human self-portrait, the outline of a human hand on the wall of an ancient cave. Then came a dizzying array of human artifacts, pottery and metalwork, weapons of the hunt, and tools to work the land. Along came the first stirrings of social organization, the building of great cities, the invention of writing, the birth of mathematics and philosophy, and eventually, well before our own times, the first stirrings of science. Given his background, Bronowski was dazzled by the regularities of nature, the precision of crystals, the predictability of planetary orbits, the constancy of the speed of light, as well as the inherited patterns of genetics as reflected in the elegant coding of the DNA molecule.

As he continued through this ambitious survey, Bronowski did not seek, and certainly did not find, even a snippet of data that would have denied our connections with the rest of the living world. At the stirring conclusion of his narrative we remained, in biological terms, just another animal. There is no need to contradict claims to the contrary, to place ourselves atop a mythical tree or contrive a wall between the human animal and our living kindred species. But Bronowski would not brook, even for a moment, those who would dispute the exceptional nature of the human species. In making that argument, he admitted we have learned much about ourselves from studies of other animals, citing the work of Konrad Lorenz on animal behavior and that of B. F. Skinner, who advanced psychology with studies on pigeons and rats. But, as he wryly noted, that was hardly the whole story. Studies by scientists like Lorenz and Skinner are relevant because:

> They tell us something about man. But they cannot tell us everything. There must be something unique about man because otherwise, evidently, the ducks would be lecturing about Konrad Lorenz, and the rats would be writing papers about B. F. Skinner.[8]

Some may see our links to the rest of the living world as ties that reduce us to the level of "mere" animals. But they are better understood as proud historical roots that place the human achievement in its proper context. Life is an experiment that has run for nearly four billion years. In its struggle to survive, it has covered the planet with unimaginable diversity, all built upon a single framework of molecular and biochemical unity, all related through ancestry, all in a single fabric of which we are but one part. That much is clear. But it is also clear that we are an exceptional part, a unique part. Other organisms have unique qualities as well, to be sure, but ours rise to another level. We are the only ones who have come to grasp the terms of our origins, the ones who understand their kinship with other forms of life, and we are the creatures, uniquely, who recognize and delight in the diversity and beauty of life.

Look again at the tree of life shown in Figure 3-2, the one in which we mammals are a tiny branch, from which the primates appear almost as an afterthought, a one-off tossed out carelessly by the machinery of genetic change and natural selection. Consider how much of the overall tree consists of microbial life, and how little is devoted to the organisms we consider most significant, the animals. Then consider the fact that the drawing overemphasizes our place in the scheme of things. Instead of the two hundred or so branches shown, there should be, at the species level, something close to ten million branches, of which we humans are just one. Does this mean, as some conclude and others fear, that "we," in the sense of humans and organisms similar to us, matter very little in the scheme of things? Not at all.

The tree of life documents pathways of exploration, the twists and turns that have characterized the evolutionary process as it found one solution after another to the challenges of survival. We are, for all our faults and frailties and self-delusions, one of those solutions, but only one. In such terms, every species is one among millions, so the mere occupancy of a spot at the rim of that expanding tree is nothing to boast about. Does that mean, however, that each and every twig is of equal importance? Only by the most naïve of assessments. In reality, as evolution carried out its ever-expanding search of the possibilities for life on Earth, on one branch, and one alone, it produced a creature with the potential not only to be self-aware, but with the intelligence to determine its own history and to reconstruct the very tree of which it is a part. At the moment this creature appeared, the character of life on Earth changed forever. It could now study and understand itself. And one of its most important findings was its relationship to the rest of the living world.

On a global level, this means that our recognition of that kinship can and should give rise to a sentiment of which no other creature is capable, and that is stewardship of the living world. We are the ones who have stepped forward from the many forms of life into consciousness and awareness. We may be an accidental species, and we are surely little different from others in physical terms, but our rising mastery of science and our awareness of the rest of life place us in a position no other species has ever occupied. It would be foolish not to appreciate that fact, and it will be tragic if we do not accept the responsibility to safeguard this planet from the excesses of our own dominance. For the very first time, life has become aware of itself, and we are the conscious vessels of that remarkable development.

CHANCE OF A LIFETIME

Critics of Darwin's work often describe his theory as "random evolution," implying not only that the outcome of the process is unpredictable, but also that it is lacking in significance, meaning, or value. If evolution in that sense were true, then our presence on this planet would become just another "random" event, of no special importance in the flow of earthly or cosmic events. In a universe where everything is random, no event, no process, no form or structure or organism would seem to be more important than any other. This view of evolution is especially troubling to those who, for reasons philosophical or religious, believe that human significance requires that our emergence on planet Earth had to be a sure thing, a predictable outcome of natural processes if not a preordained result of a Divine will.

As we saw earlier, Stephen Jay Gould called the "question of the ages" to be why humans exist. It was not, he argued, because evolution follows a predetermined pathway to perfection, not because of the innate superiority of the vertebrate body plan, and not because evolution shows any clear trend toward complexity, awareness, or intelligence. Rather, it was a matter of random happenstance. Yes, natural selection is a powerful force that can drive evolutionary change in certain directions, but the great transitions in evolution, the rise to dominance of certain major groups while others fall into extinction is largely the result of chance events. With respect to the significance of humans in the process, he wrote we should recognize that:

> *Homo sapiens* . . . is a tiny twig, born just yesterday on an enormously arborescent tree of life that would never produce the same set of branches if regrown from seed.[9]

In his popular book *Wonderful Life*, Gould emphasized his certainty regarding the improbability of our species: "Replay the tape a million times, and I doubt that anything like *Homo sapiens* would ever evolve again."[10] In a detailed sense, few scientists would dispute that point. The appearance of a nearly hairless, bipedal primate with ten fingers and ten toes in exactly human form would be most unlikely to happen twice. But Gould expanded on that point by arguing that the same considerations applied to the more general characteristic of intelligence. It would have been interesting, truly interesting, if Gould had then followed that assertion by advancing an explanation as to why the evolution of humanlike intelligence was so radically improbably. However, as Robert Wright pointed out in a critical review of Gould's book, he did no such thing:

> The real question is: Would any form of highly intelligent life have evolved if humans hadn't? Did the basic laws of natural selection make it highly probable that eventually some organism would have become conscious of itself, and even of the process that created it? Is great intelligence, generically speaking, inherent—or, at least virtually inherent—in evolution? For most of the book, Gould purports to be interested in that question. Yet he studiously avoids tackling it head on.[11]

Instead, Gould dismissed the possibility that evolution could have followed another, nonhuman, pathway to intelligence with the glib observation that so far, "All we can say is that our planet has never come close a second time."[12] Given the fact that the level of intelligence of which he speaks hadn't appeared until just two million years ago, it's difficult to understand his certainty that a second case might not develop in another form in the future along another evolutionary pathway. As Wright urged:

> Have some patience, for God's sake. The journey from a single-celled animal to a bird, to a dog, to a bear, to a chimp, took a few hundred million years. By Gould's own estimate, the Earth will probably be around another five billion years. Doesn't that leave time for a bit more action on the evolutionary front?[13]

It certainly should. Along those lines, it is worth noting that Gould based his arguments in *Wonderful Life* on groundbreaking work done with Cambrian-era fossils by a number of researchers including, most prominently, the Cambridge paleontologist Simon Conway Morris. Conway Morris's interpretation of the implications of his own findings for the course of evolution differs markedly from that of Gould.[14] In particular, he has written extensively on patterns of convergence in the fossil record, implying that evolution manages to converge on similar solutions to adaptive problems time and time again, even along unrelated lines of descent. In Conway Morris's view, this means that evolution often follows a more or less predictable pathway, driven by the need to adapt to particular ecological conditions. Putting it in other terms, as it explores adaptive space, evolution tends to find the same niches over and over again. He agrees there's plenty of chance involved, but it's not a Gould-like crapshoot.

More recently, Conway Morris has published *The Runes of Evolution*, an exhaustive tome surveying convergences across the vast expanse of the living realm. Chapter by chapter, Conway Morris describes the independent origins of scores of similar adaptations of shape, form, and biochemistry, among animals and plants with completely different evolutionary origins. These include structures for feeding, walking, swimming, and flying, as well as the organs that produce senses such as vision, smell, taste, and hearing. His point is that physical constraints, such as the laws of hydrodynamics, optics, and acoustics, limit the number of successful solutions to the problems that organisms face in particular niches, and that evolution will find those solutions time and time again.

Is intelligence just such a solution to a particular niche? Conway Morris suggests that it may be. He writes, "For Darwin, the mystery of mysteries was the origin of species, but for us it is the nature of mind."[15] So, take humans out of the picture, and for that matter, our primate relatives as well. Is there any other living group where we can see the possibility of true intelligence coming to the fore? As he suggests, consider the octopus.

The octopus is a mollusk, a group that last shared a common ancestor with the vertebrates roughly 550 million years ago. This means that any similarities between these animals and ourselves must have been achieved along completely separate evolutionary pathways. They share a body plan with squid and cuttlefish, one that is fundamentally different from the vertebrates, and yet they also display some remarkable convergences to our own physiology. They possess camera-like eyes that are remarkably like ours and have similar visual processing architecture. Lacking an internal skeleton, their muscular tentacles have no joints. Yet when grasping large objects, a wave of muscular contractions stiffens the tentacle in a way that forms a pseudo-joint very much like our elbow to leverage the weight. They learn readily, exhibit individual personality and playfulness, employ tools, and show amazing adaptability in solving problems. Videos abound on the internet showing these animals escaping from tight places and even unscrewing the lids of jars into which they have been placed.

Being invertebrates, they do not possess the myelin-coated neurons that make possible high-speed nerve impulse transmission in vertebrates, but they have solved the problem anyway. The neurons that control their jet-propulsion swimming muscles are enervated by giant nerve fibers as much as a millimeter in diameter, about a hundred times greater than a typical mammalian neuron. The huge cross-sectional area of these neurons enables action potentials to travel at very high speeds, ensuring rapid coordination between brain and muscle essential to evading predators and capturing prey. Their brains are the largest of all invertebrates and possess areas that seem to correspond to the cerebellum and hippocampus regions of the vertebrate brain.

Their genetic complexity rivals any vertebrate and contains interesting hints of the molecular basis of their complex behavior. The octopus genome contains many more protein-coding genes that the human genome (more than 33,000 according to a 2016 sequencing study), and shows a massive expansion of two gene families usually associated with the development of complex nervous systems in vertebrates.[16] These may be telling hints of what is to yet come in their evolutionary trajectories. As Conway Morris puts it, all these findings amount to clues "that stirring in the brain of the octopus is yet another approach to a universal mind."[17] The emergence of analytic intelligence and self-awareness, the neurological crown jewels we associate uniquely with the human mind, may not be a one-off, never to be repeated. Rather, as he suggests, they may be part of an amalgam of laws and principles baked into the universe from its very beginning.

If Conway Morris is right, there is a "deep structure" to life dictated by the conditions of existence. Not everything is possible in terms of physics, genetics, biochemistry, and physiology, and therefore the task of evolution, if you will, is to find those few solutions that are possible, that will work, that will stand the constant challenges of natural selection.

In this sense, there is at least an element of predictability to the evolutionary process, a set of patterns that explains the living world as something more than a random walk through creatureland. If intelligence is one of those solutions, dwelling in a niche found once by the mammals, it could be on the verge of being found again by the mollusks or another living group. And what this suggests is, as Wright put it, is that "great intelligence" may be "inherent" in evolution.

It is possible, perhaps even likely, that the appearance of humanlike intelligence is part of the deep structure of nature probed again and again by the evolutionary process. Far from being a long-odds role of the dice, the emergence of a creature who is like us in ways that matter, which is to say, with reflective, intelligent, self-awareness, was always a necessary possibility in waiting, a niche to be discovered, a creature to be realized as part of the natural world. If so, we are far more than just another random spin-off from the gears of genetic change. We embody the realization of life's ultimate potential—to know itself.

A MATTER OF THE MIND

In one of the greatest plays, the character of Hamlet describes our species as "noble in reason," "infinite in faculty," as no mere animal, but "The beauty of the world! The paragon of animals." "And yet," he continues, "to me, what is this quintessence of dust?" Not much. Certainly nothing in which one might delight, according to the Prince of Denmark. Hamlet, of course, was having a very bad day, and it was about to get worse. His foul mood, therefore, may have been understandable.

When I first read the play as an unenthusiastic high school student, I figured I knew what *quintessence* meant, since I'd heard the term *quintessential* used many times before. Snuffy's, in nearby Scotch Plains, was the quintessential steak house, Mickey Mantle was the quintessential baseball slugger, and Marilyn Monroe the quintessential movie star. Anything quintessential was the pure and perfect essence of its kind, anyone quintessential was the embodiment of perfection in their skill or craft. So, I extrapolated Hamlet's reference to mean that although we were formed from the dust, to use biblical language, we were just about as perfect as dust could be. But *quintessence*, I realized much later, implies something more subtle.

Having suffered through a couple years of Latin, I should have noticed that prefix *quint* comes from the Latin word for "fifth," so the term means "fifth essence." To the classical Western mind, the physical world was made up of four elements, or essences: fire, earth, air, and water. But to explain the substance of mind, the spiritual nature of being, a fifth essence was required, since the four essences of mere matter obviously could not account for the "infinite faculty" of human nature. Hamlet, in Shakespeare's provocative language, acknowledged this belief, only to later mock it in his graveyard speech:

> Alexander died, Alexander was buried, Alexander returneth into dust; the dust is earth; of earth we make loam; and why of that loam, whereto he was converted, might they not stop a beer-barrel?

Death, as Hamlet grimly noted, reduces that marvelous quintessence to the most mundane of earthly substances. So much for the "paragon of animals." Thou art dust, and to dust thou shalt return. This much Shakespeare understood and placed on the lips of his most memorable character for audiences to ponder.

Hamlet notwithstanding, resistance to the idea of human identity as essentially material persists and is one of the root causes of resistance to evolution. Many would look with fearful concern upon what Francis Crick called his "astonishing hypothesis":

> "You," your joys and your sorrows, your memories and your ambitions, your sense of personal identity and free will, are in fact no more than the behavior of a vast assembly of nerve cells and their associated molecules. As Lewis Carroll's Alice might have phrased it "You're nothing but a pack of neurons." This hypothesis is so alien to the ideas of most people alive today that it can truly be called astonishing.[18]

Though I might object to the minimalizing verbal twist built into Crick's language ("*nothing* but a pack of neurons"), our unimaginably huge and complex nervous system surely forms an essential part of everything we regard as human nature, since without it we literally lack a mind of any sort. Nonetheless, in chapter 5 we saw how C. S. Lewis and J. B. S. Haldane, complete opposites in terms of their theological views, were both unsettled by the notion of a mind made up of matter. In a similar way, even Alfred Russel Wallace doubted the ability of a material process like natural selection to account for the uniquely human qualities of mind displayed at the highest levels of civilization.

Many aspects of the brain remain mysterious. However, we already know that as it nearly tripled in size in a mere geologic instant, radical changes took place in its connectome, the wiring patterns of immense numbers of neurons. These changes may have produced, in the words of researchers, "a noncanonical form vital to human thought."[19] From this novel form, a truly human mind began to emerge. Neuroscientists today, regardless of their views on consciousness and free will, do not search for a nonphysical mind or a mystical explanation of the self. It's all right there, in the molecular and cellular complexity not just of the brain, but in the entirety of the human organism.

I know from many conversations with people of faith that this viewpoint remains profoundly disturbing. They doubt that everything once attributed to the human soul, its independence, creativity, and moral sense, could possibly be based in matter. To those with such concerns I commend the words of C. S. Lewis in *Mere Christianity*, who claimed it was never God's intention to fashion a purely spiritual creature. We are made of matter and have material needs. As Lewis was happy to point out, we know that God "likes matter" because, in Lewis's words, "He invented it."[20]

A large part of the aversion to a physical view of the mind surely stems from the disparaging terms with which these findings are sometimes expressed in popular media. If one, for purposes of emphasis, speaks of the brain as nothing more than a "lump of meat," it devalues the work of that organ to the point of absurdity. The old idea of dualism once invested the spirit with everything noble and the flesh with everything coarse. By carelessly describing the brain as mere flesh, the mind is stripped of its most remarkable, most human properties and portrayed as just another collection of molecules and cells. It is molecular and cellular to be sure, but it is also much more. As Marilynne Robinson has pointed out:

> It only perpetuates dualist thinking to treat the physical as if it were sufficiently described in such disparaging terms. If the mind is the activity of the brain, this only means that the brain is capable of such lofty and astonishing things that their expression has been given the names mind, soul and spirit.[21]

Indeed, many write on this issue as if the primary goal of enlightened neuroscience was to dispel the ancient concept of a spiritual soul rather than to account for the uniqueness of the human mind in scientific terms. Surely, we should live in the scientific present. The mind is composed of matter, and matter is built from elementary particles. But that does not mean that neuroscience is nothing more than particle physics, writ large. Layer upon complex layer requires one emergent discipline after another to grapple with the increasing levels of organization in molecules, biochemical pathways, cells, tissues, and finally in the nervous system itself. We may be confident that the terms and techniques of neuroscience are the proper ways to approach the brain, but that does not mean that the human mind is anything less than the wonder of the universe as we know it. It's anything but a piece of meat.

If I seem eager to defend the physical nature of mind, I am just as eager to defend the integrity of thought, the work of that physical mind.

Consider, as a case in point, Sam Harris's extreme claim that we are nothing but "biochemical puppets" steered by forces that are beyond our control. While Harris may wish to make a case against free will, against what he regards as the oppression of religion, and against pointless guilt, his arguments go much further than he seems to realize. As we have seen in some of the excesses of evolutionary psychology, it seems possible to come up with an evolutionary explanation for nearly every human behavior, from shopping to writing poetry, from styles of dress to competitive athletics. For every decision or preference or choice we make, there seems to be a "real" reason, explicable in evolutionary terms, behind our actions.

This means, for example, that when I drop a few dollars of change into the collection bucket of a charity at holiday time, my actions are dictated by a hidden Darwinian calculus of personal advantage. Although my apparent charity is directed toward total strangers, which E. O. Wilson describes as "soft-core altruism," my capacity for so doing has been produced by the "selection of individuals" over time who are susceptible to the "vagaries of cultural evolution." Why am I so willing to put a few bucks into the pot for some family other than my own? Because evolution has deceived me into thinking that this act is both appropriate and noble. How has it done so? As Wilson explains, by favoring psychological vehicles that enable "lying, pretense, and deceit, including self-deceit, because the actor is most convincing who believes that his performance is real."[22]

Apparently, only the evolutionary psychologist is able to see through this mist of self-deceit to comprehend the genuine reasons for altruism. It seems that we are unable to know our true motives, unable to command our own minds, and are so blinded by those biochemical strings that only certain scientific specialists can explain them to us. One might be left to wonder, then, what special gifts the scientist possesses that enable her to break those strings and discern true motives for the choices and actions of others, while never being bound by such strings herself. And, as Hamlet might say, there's the rub.

This profoundly determinist view of human thought and action threatens science itself because it undermines the most far-reaching of all human gifts, the power of reason. If evolution shaped our minds, as complex and powerful as they may be, to meet only the demands of selection, survival, and reproduction, and if we explain all the mind's actions accordingly, then reason itself is suspect. It means that the explanations we think we provide for natural events are dictated in ways we cannot control. And it means that when we test those hypotheses, however rigorously, we are applying standards dictated not by impartial tests of truth, but by deeper vehicles of self-deceit carried forward over evolutionary time. Try as we might, we cannot escape those biochemical strings.

Psychologist Paul Bloom called this tendency to explain all of human behavior in evolutionary terms a "war on reason," noting "Aristotle's definition of man as a rational animal has recently taken quite a beating."[23]

His target in a 2013 essay published in *The Atlantic* was the idea, prominent in many intellectual circles, that "the neural basis of mental life suggests that rational deliberation and free choice are illusions." If they are indeed illusions, of course, then deliberations to reach such conclusions are illusions as well, leaving us with no basis to say anything authentic about ourselves or the world in which we live. He wrote:

> Yes, we are physical beings, and yes, we are continually swayed by factors beyond our control. But as Aristotle recognized long ago, what's so interesting about us is our capacity for reason, which reigns over all. If you miss this, you miss almost everything that matters.

Science is surely one of the things that matters, and that includes the science of evolution. If we take the genuine reality of human evolution to mean that everything about us has an explanation only in terms of selective advantage, we miss nearly everything that makes our species unique. What is truly remarkable, to reword Haldane's concerns about the physical nature of the brain, is that a mind made up of atoms was able to discover the atom. It is that a creature composed of cells was able to discover, dissect, and understand the cell. And finally, that an animal produced by evolution could identify that very process, to understand the marks that descent with modification left on body and mind, and then to rise far above the demands of mere survival. Evolution does not undermine our humanity, our capacity for reason, or our science. It is, in fact the foundation of each. We have become the reasoning animals we are because we are the products of evolution.

A THEORY OF EVERYTHING?

One of the oft-stated goals of physics is to develop a "theory of everything," a small number of interlocking theoretical principles that would explain the properties of matter and energy, as well as all the fundamental forces of nature. In the words of Stephen Hawking, such a theory "would be the ultimate triumph of human reason—for then we should know the mind of God."[24]

Hawking's "God," of course, is not spiritual but metaphorical and refers not to a supreme being but to a single set of overarching principles that might govern the physical world. While physicists have not yet arrived at such a theory, reaching it remains the goal of many researchers seeking to grasp the big picture of existence. In biology, one might well say that evolution, updated by molecular genetics and genomics, is already our theory of everything. Immanuel Kant once wrote: "There will never be a Newton for a blade of grass." But as philosopher Michael Ruse has pointed out, "There was a Newton of the blade of grass and his name was Charles Darwin."[25] Evolution is the central organizing and explanatory principle of biology, which is why biologists tirelessly repeat the aphorism that *nothing in biology makes sense except in the light of evolution.*[26]

Naturally, I concur. But does its ability to explain human biology also mean that evolution tells us everything we need to know about human culture? As we have seen, David Sloan Wilson is confident that the "light of evolution" extends well beyond mere biology to encompass fields such as anthropology, art, economics, politics, psychology, and history.[27] In short, evolution will soon impose a kind of cultural imperialism to which the humanities and social sciences must submit or wither and die. All true explanations of human creativity, even in the arts, are now Darwinian.

Enthusiasts of this sort have not been shy about subjecting literature, music, and fine art to purely Darwinian interpretations. Examples abound. They include Brian Boyd's attempt to explain literature, *On the Origin of Stories: Evolution, Cognition, and Fiction*, Philip Ball's *The Music Instinct*, and *The Art Instinct*, by Denis Dutton, which is subtitled *Beauty, Pleasure, and Human Evolution*. Dutton's work is representative of this genre and bases its analysis of art on three basic principles, each of which he takes as a given: (1) Art is an adaptive cultural phenomenon, which means that it has value in terms of survival and reproductive success. (2) This not only explains why we do art, but how we should judge it. (3) Since art is pan-cultural, there must be an "art instinct" shaped by evolution and explicable in evolutionary terms.

If Dutton's only point was that humans possess an aesthetic sense that leads us to create, appreciate, and value visual images, who could argue? But his claims are far more specific. For example, he notes surveys showing that people, especially young people, seem to prefer paintings that depict landscapes containing, people, animals, and water. He explains this preference by claiming that such paintings share "an intriguing relationship to the African savannas and other landscape forms hospitable for human evolution."[28] To Dutton, such preferences are not byproducts or spandrels resulting from a broad and general aesthetic sense. Rather, he writes, "It is wrong, however, to regard these as by-products of prehistoric impulses or emotions: rather, they directly address and satisfy ancient, persistent interests and longings."[29] This formulation is remarkably similar to E. O. Wilson's statement about the headquarters of the John Deere corporation, although the African savannah, as I have noted, surely did not include fountains, manicured shrubbery, and picturesque walkways. Nor, I would note, is the savannah anything like the much beloved European landscapes of Corot or the paintings of the Hudson River school. One is left to wonder just how robust that aesthetic link to our specific ancestral past might actually be.

Compounding these difficulties is a lack of any explanation for great art depicting profoundly nonidyllic landscapes. Take for example, Paul Nash's bleak 1918 canvas *We Are Making a New World*. Completed near the end of World War I, Nash's landscape is the artistic contradiction of nearly every criterion by which Dutton would explain and judge a painting. There are no animals, no people, and there is certainly no water. Instead, we confront a haunting scene of pockmarked mud and lifeless, broken trees, with not a living thing in sight. The landscape may not be beautiful, but it is powerful and compelling nonetheless in its depiction of the "world" fashioned by the great powers as a result of four years of war. Dutton's simple explanations of nostalgia for our ancient home turf in Africa cannot even approach the question of why Nash's work qualifies as art. One is left to puzzle how he would approach the terror inherent in Picasso's *Guernica*, or the biting, highly targeted political sarcasm of Diego Rivera's *Glorious Victory*.

Even more suspect is the explanation that Dutton and others such as Geoffrey Miller offer for the adaptive value of the arts broadly, including storytelling, music, and visual art. According to Dutton, being able to speak or sing or paint in a way that others find attractive gives you a leg up in the evolutionary arms race. He writes: "Speech performances, especially artistic speech performances, are Darwinian fitness indicators: ways of judging the wit, originality, or general cleverness of a person." And further, "our admiration of skill and virtuosity itself is an adaptation derived from sexual selection off the back of natural selection."[30] Why do such "fitness indicators" matter? Because they are of prime value in mate selection. In other words, art is created to help a man find a mate and get lucky with her. My placement here of the male and female pronouns is not accidental. Dutton and Miller both emphasize that by and large, the guys make the art, and the gals do the choosing. That, presumably, is why most artists, composers, writers, and even comedians are male. If you find this formulation more than just a tiny bit sexist, welcome to the club.

Dutton, incidentally, also uses the principle of sexual selection to explain why we value an original painting far more than a copy or forgery, even if the technical skill of the imitation work is every bit the equal of the original. It's because the forger is actively undermining the evolutionary value of the artwork by assigning reproductive fitness to a person (the forger) not actually responsible for the artistry of the original. Presumably, evolution has also programmed us to recognize "cheats" in the mating game and to punish them appropriately so the integrity of art as a sexual ornament is preserved. But one could make an even more powerful nonsexual case that evolutionary pressures on the instincts that underlie social interactions would have produced strong negative reactions to individuals who lied, cheated, or stole credit from others. Such an explanation would apply to all human interactions, and not just art, and therefore would be the subject of far greater selective pressure.

As we examine Dutton's claims, it's also worth noting that detailed information regarding the actual human and social environment of the Pleistocene, which is when selection for these traits is supposed to have taken place, is lacking. Anthony Gottlieb, who reviewed Dutton's book for the *New York Times*, was quick to point this out:

We know so little about the environment of our Pleistocene ancestors, what they were like, and how they lived, that almost any hypothesis about which strategies might have helped them to reproduce, and thus let their characteristics ripple through the gene pool, is bound to be highly speculative.[31]

"Highly speculative" is exactly the right term to apply to Dutton's fitness indicator explanation for the art instinct. It could also be applied to our love of poetry, music, drama, and literature. And why stop there? Once we've decided that art can be explained by assuming that artsy guys get the girls, why not extend that explanation to sports? Doesn't the football hero always get to date the homecoming queen? And maybe we play chess just to impress upon onlookers that we carry a basket of great genes for cognitive intelligence? Are sex and evolutionary nostalgia really the explanation for every aspect of human culture? As a book reviewer for the *Washington Post* wrote, "arguing that the sex lives of poets explain the origins of poetry makes about as much sense as using the bedroom exploits of Wilt Chamberlain to construct a biological explanation of basketball."[32]

Even if we accept the delightfully risqué underpinnings of Dutton's origin story, it fails to explain what really matters in the study of art. If evolution shaped a universal set of aesthetic preferences, why do the representational styles of art differ so widely from one culture to another? What accounts for the "evolution" of different artistic schools, from classical to impressionist to cubist to modern to postmodern and beyond? How could an artist like Picasso begin a successful career producing clearly realistic works, and then move on to cubist and finally highly abstract works and yet find popular acclaim every step along the way? Above all, how are we to judge art, how do we distinguish greatness? Why do we take a single work of portraiture, the *Mona Lisa*, and elevate it above so many other works that likewise depict a human figure set against a natural background? As one critic observed, while a theory like Dutton's art instinct can explain the sentimentality of mediocre art, that's all it can explain: "When everything in the Museum of Modern Art violates your theory of aesthetics, then it might be worth revising the theory."[33]

The notion that human culture and aesthetics can be completely subsumed by biology grossly overstates the reach of science and dramatically undervalues the significance and complexity of human thought.

Evolution explains the origin of species, ourselves included, but it is simply not a cultural or social theory of everything.

As a critic of Dutton pointed out, explaining works of art in terms of evolved reactions to ancient selective pressures rather than as "inspired creations" undermines "the meaning and significance of human art."[34] Attempts to do the same with music, literature, drama, and the other creative arts have done likewise. And they have done little to illustrate the true nature of the human genius that lies behind such creativity.

Evolutionary biology will not take over the humanities, nor will it reduce the artistic and cultural achievements of human society to by-products of a rush to survive and reproduce. We are creatures capable of the fugues of Bach, the verses of Yeats, the stories of Twain, the creations of Dalí, and, for that matter, the mathematics of Gödel, Ramanujan, and Turing. But these masterworks emerged *from* the mental platform that evolution fashioned, not as the product of evolution itself. And the human creativity behind each speaks not to a set of rigid tendencies programmed by natural selection, but to the limitless possibilities that evolution put within our reach.

CENTER STAGE

An interesting debate is going on right now among scientists, and we are its focus. Earth's three most recent geologic eras are the Paleozoic ("old life"), which lasted until the rise of the dinosaurs, the Mesozoic ("middle life"), which ended with the last of the great dinosaurs, and the Cenozoic ("new life"), which marks the rise of the mammals. We inhabit that most recent era, the Cenozoic, or so the textbooks would tell us. However, there is a serious and well-supported movement among geologists, paleontologists, and other scientists to recognize and name a fourth era—the *Anthropocene*, literally, the "human era."

While academic scientists debate whether we are in such an age and when it may have begun, this much is clear: we live at a time when our own species has taken mastery of planet Earth in a way that affects not only its great global cycles of matter and energy, but the very existence of other forms of life. We are not only the planet's dominant mammals, but we have taken nearly three quarters of the Earth's land area[35] to our own use. We have changed the composition of the atmosphere, the acidity of the oceans, and the level of the seas. Sadly, we may be about the preside over the Earth's sixth mass extinction, and many of us actively wonder whether our societies will gain the wisdom to understand the consequences of our actions in time to reverse these and other trends. The answer to that question is in doubt. But for better or for worse, the blue planet, third from the sun, has become truly ours in every sense that matters.

Catastrophic changes have happened before on planet Earth, and at times living organisms have caused them. Two billion years ago the first photosynthesizing organisms captured the energy of sunlight for their own benefit and spewed into the atmosphere a waste product that no doubt killed millions of other living things. That waste product was oxygen, and its insertion into our atmosphere was fatal for those organisms that could not deal with its highly reactive chemistry. What's different this time is that we, the drivers of these changes, know they're taking place. We are aware not only of what's happening to our planet, but also of the chemical, physical, and economic reasons behind those changes. This time, it truly is different, and we are the reason why. Rather than the age of man, *Anthropos*, perhaps we should call ours the age of awareness. For the first time, a form of life has become conscious in a way that allows it to understand the consequences of its actions and to make judgments as to whether it should continue on its present course.

In no small measure, science itself is a profound consequence of such awareness. The ability, as well as the uniquely human motivation, to interrogate nature is at the heart of the scientific enterprise. Evolution has brought one organism, and one organism only, to that point. I would argue that should be cause for celebration, not for despair.

Nonetheless, despair is woven throughout Marilynne Robinson's essays in *The Death of Adam*, where she laments the passing of a story that once defined humanity's place in the world. In her view, the Darwinian narrative has cheapened human life, brought on a ruthless commercialism that regards market success as the sole arbiter of human values, and redefined morality, art, literature, and faith as the meaningless residues of natural selection. As she puts it, with characteristic conviction and elegance:

I want to overhear passionate arguments about what we are and
what we are doing and what we ought to do. I want to feel that
art is an utterance made in good faith by one human being to
another. I want to believe there are geniuses scheming to astonish
the rest of us, just for the pleasure of it. I miss civilization, and I
want it back.[36]

Robinson's thoughtful essays are representative of those who accept the
realities of evolution, who cannot abide the pseudoscience of "young
Earth" creationism but nonetheless feel deeply troubled by the views of
human nature that have taken Adam's place in our culture. One feels
the same sense of helpless sadness in Ian McEwan's *Saturday*, a novel
that found its characters deeply moved by laments of the modern in
Matthew Arnold's "Dover Beach."

As Robinson is well aware, we do not get to choose the natural
history of human origins any more than we get to pick the value of
Planck's constant, the dipole moment of a water molecule, or the speed
of light. The natural world is out there, real and genuine, and it is ours
to discover, not to construct. What is very much ours, however, is the
possibility of deciding what to make of these discoveries. If Robinson
really hopes to overhear passionate arguments about who we are and
what we ought to do, I would suggest that she try hanging around for
drinks in the evening after a few provocative talks at a scientific meeting. She might be surprised at the breadth of such discussions and how
often they address exactly the points she feels to be so neglected.

Meaning is not a quality inherent in either matter or energy. A carbon atom merely is. It is not *for* anything, it does not *mean* anything, it
has no *purpose*; it is itself and nothing more. Meaning is something that
we assign, something that comes from individual conscious reflection
and thought. Many years ago, I gave my wife-to-be a small crystal of
carbon atoms. As she would tell you, given the wages of a graduate student, it was a *very* small crystal. But its meaning, to her and to me, was
every bit as real as the physical reality of the atoms themselves.

The reality of our existence is that we are creatures of matter, like
every other living thing. Like other living things, we have an evolutionary history rooted in struggle and pain, but also in persistence, survival,

and success. As we come to understand those facts, what meaning should we assign to them? What should we make of the realization that our thoughts are chemical, rooted in the flows of ions and neurotransmitters between cells, that we inherit behavioral instincts honed by natural selection, and that many of our immediate perceptions of the natural world are flawed by the limitations of our senses? I suppose we could shrug and accept what E. O. Wilson tells us all this means:

> The unfolding of history is obedient only to the general laws of the Universe. Each event is random yet alters the probability of later events. During organic evolution, for example, the origin of one adaptation by natural selection makes the origin of certain other adaptations more likely. This concept of meaning, insofar as it illuminates humanity and the rest of life, is the world-view of science.[37]

Wilson does admit to a second level of meaning, which is intentionality behind the actions of a living thing, such as a spider spinning its web. The meaning of the web is to catch a fly. In the case of humans, we not only undertake intentional actions for a particular purpose, but unlike the spider, we have the capacity for informed, conscious decision making:

> Every decision made by a human being has meaning in the first, intentional sense. But the capacity to decide, and how and why the capacity came into being, and the consequences that followed, are the broader, science-based meaning of human existence.[38]

As well intentioned as Wilson's stab at meaning might be, in all fairness, the capacities to decide and to evaluate consequences are *properties* of human life, not its meaning. If that is all science can tell us about existence, then it is pretty thin stuff indeed, and I am sure Robinson and McEwan would agree. Furthermore, if we accept as the verdict of evolution that our free will and decision-making capacities are themselves illusions, then nothing seems to be left of even Wilson's paltry formulation. Once it was said that we would know the truth and the truth would make us free. But these truths, it seems, have left us in a prison of our own making, where even the hope to find meaning has been banished.

I began this book by noting that the associations of many with the idea of human evolution are grim—describing accidental creatures whose existence is pointless, dwelling in a world of illusions carved out by the demands of natural selection, and whose actions are robotically dictated by the deterministic neural chemistry of reflex and reaction. Ironically, this view of the implications of human evolution finds its most passionate adherents among those who take the most extreme views of the meaning of the theory of evolution—embracing it on one side, rejecting it outright on the other. In this way, the antievolutionists have found their best allies among those who argue for the most extreme—and most dehumanizing—view of the significance of the evolutionary process.

There is another viewpoint, however, and it is the one I have sought to develop in these pages. If we truly accept the validity of science, as I think we must to say anything about the natural world, our view of the human animal cannot help but acknowledge its exceptional nature. Yes. There are elements of our ancestry that depended, in a sense, upon "random" chance. But it is also true that we emerged in a way dictated by the laws of physics and chemistry and by conditions that have prevailed over eons of change on this planet. We are children of the Earth, but we are more. We are sons and daughters of the Cosmos in every sense that matters. To view our existence as nothing more than an insignificant accident in an obscure corner of the universe is to discount the very science that makes it possible for us to draw that conclusion in the first place. We are the best and brightest things we know, and it is for us to set the terms by which we address existence and define the meaning of our lives.

As Marilynne Robinson observed, biology tends to disdain the idea of evolutionary progress. Life on this planet did indeed begin simply as a matter of chemical reactions. It then gave rise to living cells and ultimately to multicellular organisms with higher levels of organization and complexity. However, since evolution may also proceed in ways that simplify and reduce complexity, there is no inherent direction, other than the endless and ever-changing exploration of adaptive space. But in that very process of exploration, evolution found a pathway leading to creatures with levels of consciousness, awareness, creativity, and intelligence unique among all its creations. We are those creatures, and in fulfilling the philosopher's command to know ourselves, it is only proper to ask what is meant by our presence in the living world. In Sagan's characteristically grand answer to that question, "we are the local embodiment of a Cosmos grown to self-awareness." The universe has begun to recognize and study itself, and we are the living agents of that progression to conscious, scientific awareness.

In the Book of Genesis, God assigned to Adam the task of naming all the animals,[39] a labor he apparently accomplished by skipping the insects, since he finished in a single afternoon. Our work is going to take longer than Adam's, because we have assigned ourselves the goal of understanding all of existence, from the very small to the whole of the universe. If we seek meaning to our lives, there it is. As Bronowski wrote, "Knowledge is our destiny. Self-knowledge, at last bringing together the experience of the arts and the explanations of science, waits ahead of us."[40]

Thomas Nagel has expressed impatience at what he regards as the failings of materialist science to produce a full accounting for the evolution and workings of the phenomenon of consciousness. That science, which includes evolution, is incomplete in his view because it cannot account for a "cosmic predisposition" to the formation of life and consciousness.[41] What is needed, in his view, is an explanation of the tendency for life to form in the Cosmos, a tendency he thinks cannot be explained "by the nonteleological laws of physics and chemistry."[42] While I do not agree with Nagel on this point, the question of teleology, of a goal or direction to natural history and to evolution, remains.

It would be foolish to declare humanity the endpoint of evolution, the *telos*, or the goal, of history. Evolution goes on, and it continues to change the living world, ourselves included. But it is more than fair to ask whether the universe itself shows a tendency toward consciousness and self-awareness. We are, of course, a material part of that universe in every sense. We are conscious and self-aware. And, therefore, at least a small part of that universe has indeed reached that goal, if we may be bold enough to call it a goal. The nature of the universe's journey to consciousness and its embodiment in human nature matters, and it matters in the deepest sense. It assigns meaning to our view of the world, our relationship with nature, and even the ethical approaches we take in forming societies and dealing with other species. It is important, vitally important, that we get the details of that journey right, that we tell the truth about our biological origins, and that we draw the wisdom of experience from an understanding of our evolutionary past.

Our biological heritage is merely the beginning of what we can be, not the end of it. The understanding that we have our roots in the process of natural selection explains how our brains were shaped, but that does not undermine the independence of human nature or deny the reality of human knowledge and achievement. Evolution may explain the human need for art, music, religion, and even science, but it cannot explain those disciplines away. Each exists, in its highest form, as an expression of the best humanity can offer in making sense of this remarkable world.

We hold in our hands the power to determine whether we will turn out to be the heroes of our lives or if that task will fall to some other creature at some other time. That, I would suggest, is something that should fill us not only with a sense of responsibility but also with pride, and it should instill within us a sense of the grand and cosmic nature of every moment of the human experiment. We stand at center stage in the pageant of life. It is our time to shine, and the role we will play in the history of our planet and of the Cosmos itself matters beyond all measure.

We may have started as just one branch on Darwin's tangled riverside bank of life, but we are the branch that emerged to make sense of it all. Some may take a certain mild satisfaction from reflecting upon this view of life, but I will argue that the more appropriate emotions are joy and delight. Joy that we are approaching a genuine understanding of the world in which we live, and delight at being perhaps the very first stirrings of true consciousness in the vastness of the Cosmos. Far from diminishing us, knowing the details of Adam's journey ennobles each of us as a carrier of something truly precious—the genetic, biological, and cultural heritage of life itself. Evolution describes not the death of Adam, but his triumph. That is the great truth of our story.

The Chromosome 2 Fusion Site

A Technical Appendix

The human chromosome 2 fusion site was described in general terms in chapter 2. For those who might be interested in a more technical treatment of the evidence, I've included this appendix.

In 1982, Jorge Yunis and Om Prakash at the University of Minnesota Medical School decided to take a comparative look at the chromosomes of humans, chimpanzees, gorillas, and orangutans. Spread out and stained to make their internal banding patterns visible, the scientists took photographs of each set of chromosomes under a light microscope, and then lined each chromosome up with the corresponding one from the other three species (see Figure 2-5). They noted that their results demonstrated a "remarkable similarity in the banding patterns of most chromosomes," implying a common ancestor. From those similarities, they attempted to reconstruct the chromosomal organization of that common ancestor. Humans have 23 pairs of chromosomes,[1] and these scientists proposed that 18 of those 23 pairs were "virtually identical" between modern humans and that common hominoid ancestor. The remaining pairs were "slightly different." Several showed inversions in which portions of a chromosome in one species had been flipped around to a reverse orientation in another. But the most obvious and striking difference was found in the composition of human chromosome number 2.

At first, there didn't seem to be a match for it among the chromosomes of the other great apes. But then they looked a little closer and discovered that each of its two halves did indeed match one of the chromosomes of the other species. As they noted, the two halves of human chromosome 2 correspond to two separate chromosomes in the chimpanzee (as well as in gorillas and orangutans), indicating that our second chromosome was formed from an end-to-end fusion of those chromosomes. In case the printed photomicrographs were not clear enough, Yunis and Prakash included a diagram of banding patterns to illustrate this point:

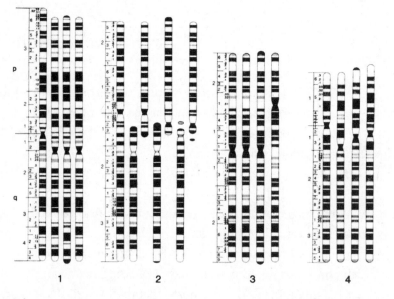

Schematic representation of the banding patterns in chromosomes 1 through 4 of humans, chimpanzees, gorillas, and orangutans. Note that the two halves of human chromosome 2 line up with chromosomes that are still separate in the other primates.

As the authors wrote, "Evidence for a common ancestor of man and chimpanzee also comes from chromosome 2." Why? Because the most straightforward explanation for the origin of our second chromosome is an end-to-end fusion of two chromosomes that are still separate in our primate relatives. The tips of chromosomes are known as "telomeres," so they called this a "telomeric fusion." More on that in a bit.

In 1982, the best technique that Yunis and Prakash (or any other re-searcher) could use to compare the chromosomes of these species was light microscopy. Today, however, we have complete genome DNA sequences for these species, and can do a much more sophisticated comparison.

One can, for example, go to any of the online genome database websites and do a direct analysis of similarities and differences between these chromosomes. One such site is Ensembl, which accesses databases maintained by several research organizations. A quick trip to www.en-semble.org allows any user to select the human genome, then view a karyotype that displays all 23 chromosomes, and finally to click on chromosome 2. The database allows a user to make sophisticated com-parisons of genetic "synteny" between human chromosomes and those of other species. Synteny is the degree to which the order of genes and DNA sequences line up between the chromosomes of two species. Ask the database to analyze the synteny displayed by human chromosome 2 with chromosomes of the chimpanzee genome, the orangutan, or the gorilla, and an image like this will appear:

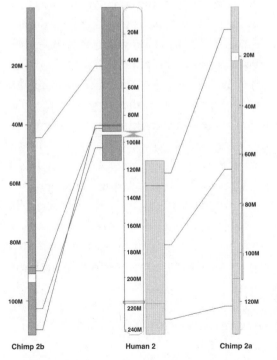

Ensembl comparison demon-strating synteny between human chromosome 2 as compared to chimpanzee chromosomes 2A and 2B.

While there is a tiny piece that matches up with fragments of another chromosome, nearly all of human chromosome 2 is a match, just as Yunis and Prakash inferred, for two chromosomes in these great ape genomes, now known as 2A and 2B. At some point in the past, these two chromosomes became fused together to produce human chromosome 2, and the synteny, the order of genes, on either side of the fusion site has been largely preserved. In fact, if one delves deeper into the chromosome maps at Ensembl, one can see that the order of specific genes on either side of the fusion has indeed been preserved. That order is:

> IL1RN – PSD4 – PAX8 – CBWD2 – Fusion –
> RABL2A – SLC35F5 – ACTR3 – DPP10

The genes to the left of the fusion site are found on one end of chimpanzee chromosome 2A, while those on the right match chimpanzee 2B. As the comparison shows, we do indeed possess a chromosome that was indeed formed by the fusion of two chromosomes still separate in our primate relatives. Yunis and Prakash were correct.

Now, what about the fusion site itself? A fusion between two chromosomes must involve the ends, the "telomeres" of both chromosomes. The ends of linear DNA molecules, like those in human chromosomes, are notoriously difficult to replicate, something a cell has to do before it can divide into two daughter cells. There are technical reasons for this, due to the way in which DNA polymerase, the enzyme that copies DNA, works as it moves along a chromosome. Because of these difficulties, the DNA sequences in telomere regions are replicated by a different enzyme known as telomerase. Telomerase doesn't copy DNA the way that DNA polymerase does. Rather, it strings together long repeating stretches of DNA that don't actually code for anything. In humans and other primates, telomerase makes hundreds of copies of a 6-base sequence, TTAGGG. So, if you were to read the bases in a human telomere region, you'd see that sequence repeated over and over again, something like this:

...TTAGGGTTAGGGTTAGGGTTAGGGTTAGGG...

Actually, since DNA is double-stranded, if you were to look at both strands together, such a region would look like this:

·   ...TTAGGGTTAGGGTTAGGGTTAGGGTTAGGG...
    ...AATCCCAATCCCAATCCCAATCCCAATCCC...

If human chromosome 2 had indeed been formed by a fusion of two chromosomes, two groups of telomere sequences should have been squashed together right at the point of fusion, one from each of the original chromosomes. In 1982 (the time of the Yunis and Prakash study) there was no way to check this. But once the human genome, including chromosome 2, had been sequenced via the Human Genome Project, it was possible to take a look. Today we can read the DNA base sequence of nearly the entire chromosome and look for those telomere repeat sequences where they wouldn't normally be found, right in the middle of the chromosome. Sure enough, there is a region in the middle of the chromosome with roughly 150 of these telomere repeats, right at the fusion point. In fact, it's even possible to pinpoint the exact base where the fusion took place: 113,602,928 bases from the end of the chromosome according to the most recent human DNA sequence. But can we tell for sure? Well, the degree of synteny between chromosome 2 and chimpanzee chromosomes 2A and 2B on either side of the fusion site should be evidence enough.

However, there is a second sign that is even more compelling in terms of evidence that was provided in 1991 by a research group at Yale.[2] If a break occurred in the telomere regions of two chromosomes to fuse together, there should be a dramatic change in the pattern of DNA bases right at the point of fusion. Remember when we saw that the human telomere repeat pattern was TTAGGG? Well, if two telomeres had been fused together, the sequence at the fusion point should change to match the opposite strand of DNA, specifically, it should read CCCTAA. That's because at the point of fusion one chromosome is, in effect, flipped around 180 degrees when it lines up with the other. The Yale group discovered that there is exactly such a point right at the fusion site:

...TTAGGGG-TTAGGG-TTAG-CTAA-CCCTAA-CCCTAA...

The switchover from the TTAGGG pattern to the CCCTAA is found in the midst of no fewer than 158 repeats of the telomere sequence. What started out as nothing more than a low-resolution snapshot of whole chromosomes has now been studied right down to the molecular level—and the fusion story has been confirmed every step of the way.

The chromosome 2 story is simple, clear, understandable, and therefore powerfully persuasive. But for those in denial of human evolution, it presents a serious problem. From the day I used it as part of my testimony in the landmark *Kitzmiller v. Dover* trial in 2005,[3] critics of evolution been trying to come up with reasons to reject this narrative and undermine the science behind it.

One such argument is that there are too few copies of that 6-base telomere repeat at the fusion site. A typical human telomere may have more than a thousand such repeats, while the fusion site has fewer than 200. But, as it turns out, that's actually to be expected, since one of the functions of the telomere is to prevent chromosome fusion. Research has shown that when the number of telomere repeats is reduced, fusion becomes much more common.[4] Therefore, a relatively reduced number of repeats at the chromosome 2 site is actually what should be expected for two chromosomes that fused because of the loss of DNA at their telomeres.

Along the same lines, some critics have pointed out that the repeats on chromosome 2 don't perfectly match the TTAGGG pattern. Of the 158 repeats around the supposed fusion site, many if not most are "degenerate." Some are 5 bases long, others 7. Some read TAGGGG or GTAGGG or TTGGGGG instead of TTAGGG. But that, too, makes perfect sense. Telomerase, the enzyme that produces those beautifully uniform 6-base repeats, only works at the tips of chromosomes. Once two chromosomes fuse and their telomeres wind up in the middle of the chromosome, all sorts of forces take over that change and degrade the repeats. These repeats produce replication errors, generating mutations, and recombination between chromosomes causes changes in the number of bases per repeat. Since the fusion occurred several million years ago, a degenerate repeat pattern is *exactly* what we would expect to find today.

Recently, a paper published by the Institute for Creation Research (ICR) has argued[5] that an active gene is located at the so-called fusion site, and therefore the whole chromosome 2 story has been falsified. The basis for this claim is a pseudogene known as DDX11L2, located next to the fusion site. According to some genome databases, this gene is transcribed into an RNA molecule that actually spans the alleged fusion site. Therefore, this isn't a fusion site at all, but is part of a complex, working gene that was "designed" as part of the human genome.

This is an interesting criticism, but ultimately it backfires in a most revealing way. My own reading of current genome databases indicates that this particular gene doesn't span the fusion site, but is located just next to it. It's a pseudogene, of course, and its function, if any, is still not known. In some databases, DDX11L2 does indeed show one reported RNA transcript that continues past the fusion site. But the databases show something else as well. Specifically, as a 2009 research paper[6] pointed out, DDX11L2 is part of a family of 18 pseudogenes scattered around the human genome. Can you guess where 17 out of those 18 are located? Every one is found right next to the *telomeres* of other chromosomes. The lone exception is the one located at the chromosome 2 fusion site. This is profoundly compelling evidence that it, too, was once part of a telomere, just as the fusion story demands.

It gets even better. Next to each and every one of the DDX11L pseudogenes is another gene, abbreviated WASH,[7] located on the side of DDX11L away from the telomere. There's also a WASH family gene in the middle of chromosome 2, next to the DDX11L2 pseudogene, and located on the side away from the degenerate telomere sequences in the fusion site, just as you might expect. Finally, in every one of its 18 locations in the human genome, the DDX11L pseudogene is transcribed in a direction away from the telomere, while the WASH gene is transcribed in a direction toward the telomere. The same is true for the copies of these genes located near the chromosome 2 fusion site. In fact, every landmark surrounding the DDX11L2 gene suggests that it was once located next to a chromosome telomere.

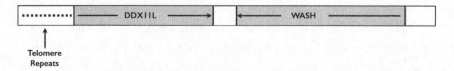

Telomere
Repeats

All members of the DDX11L pseudogene family are flanked by telomere repeat sequences on one side, and by a member of the WASH gene family on the other side, transcribed in the opposite direction, as shown above. This includes DDX11L2, the pseudogene located at the chromosome 2 fusion site, confirming its identity as derving from an ancestral telomere region.

The ICR paper, citing the location of the DDX11L2 pseudogene, argued "the evolutionary idea of the chromosome two fusion in humans should be completely abandoned." The breathtaking irony of such a statement is that the very observations cited in that manuscript actually strengthen the case for fusion. By pointing out that the "alleged" site contains a pair of genes found adjacent to telomeres *everywhere* else in the human genome the paper has actually done yeoman service in confirming the chromosome 2 story.

Finally, some critics have argued that chromosomal fusions are simply impossible, or they would cause infertility to the point where such individuals would be unable to produce offspring. So, despite the molecular evidence, this just couldn't have happened. When I first heard this argument, I referred critics to the fact that fusions very much like this are found in many animals, and are well understood. They are quite common in mice, for example, and have been well studied in horses. Domestic horses, for example, have 64 chromosomes, while the last remaining populations of truly wild Asian horses (known as Przewalski's horse) has 66, the most of any other species of equid. Donkeys have just 62 chromosomes, and the differences between these species, all derived from a common ancestor, can be traced to chromosome fusions and translocations.

Even more compelling, however, is some recent genetic data from our own species. While it's entirely possible that chromosomal fusions might initially affect fertility, that doesn't seem to hold for the long term. A 2013 report out of China described a healthy 25-year-old man with just 44 chromosomes. That scientific report also includes a photomicrograph of this man's karyotype, which shows a large chromosome resulting from the fusion between chromosomes 14 and 15.[8] Supporting this view is a Spanish study two decades earlier that had described a fusion between chromosomes 13 and 14 in three members of a family that were otherwise perfectly healthy.[9] The Spanish study suggested that both parents of a family with 6 children carried the fusion between chromosomes 13 and 14. Three of their children inherited a copy of the fused chromosome from both parents, reducing their chromosome number to 44.

The fact that similar fusions occur today in our species demonstrates that this last, desperate objection to fusion is clearly not valid. Nonetheless, denial continues, and the stridency of many objections to the narrative is striking. More than anything else, this may speak to how much the critics of evolution fear the effectiveness of the chromosome 2 evidence. I suppose we should regard that as a kind of compliment to the efforts of many scientists and science writers who have worked to spread this information. But it is also testament to the passion with which narrative of human evolution continues to be resisted.

Notes

Prologue: Our Story

1. Books written about the *Kitzmiller v. Dover* trial include: *40 Days and 40 Nights*, by Matthew Chapman; *Monkey Girl*, by Edward Humes; *The Devil in Dover*, by Lauri Lebo; and *The Battle Over the Meaning of Everything*, by Gordie Slack. The trial was also featured in two television programs, one by the BBC entitled "A War on Science." *NOVA*, the PBS science series, presented a two-hour analysis of the trial with the title "Judgment Day."
2. Harris, *Free Will*, 13.
3. S. Pinker, "Science is not your enemy," *The New Republic*, August 6, 2013.

CHAPTER 1: Grandeur

1. Darwin, *On the Origin of Species*, 759. Note: this page reference is taken from a collection of books by Darwin edited by E. O. Wilson, *From So Simple a Beginning* (New York: Norton, 2009).
2. Ibid., 760.
3. Ibid.
4. Ibid.
5. The cartoon appeared on the cover of *Punch's Almanac* for 1882. It was published on December 6, 1881.
6. McEwan, *Saturday*, 56.
7. Ibid.

8. Shapiro, *Trying Biology: The Scopes Trial, Textbooks, and the Antievolution Movement in American Schools.*

9. This quotation is taken from the Wikipedia article on *Inherit the Wind*. Attributed to a widely quoted 1996 *Newsday* story on the Broadway revival of the play.

10. That technicality was that the judge had assessed a fine of $100, while Tennessee law at the time required any fine in excess of $50 to be assessed only by a jury. Therefore, while the judge set the fine and the conviction aside, he did not overturn the verdict and certainly did not find in favor of Scopes.

11. The poll was sponsored by the University of Texas and the *Texas Tribune*. It was conducted in the winter of 2010, and involved 800 registered voters. The poll results I cited here were accessed online at: https://texaspolitics.utexas.edu/sites/texaspolitics.utexas.edu/files/201002-summary-all.pdf.

12. Incidentally, the phrase "entirely through natural selection" is not one that I or most scientists would support. There are many forces at work in shaping evolution, and natural selection is only one of them. Much more on that later.

13. Shermer, *In Darwin's Shadow*, 220.

14. Quotations in this paragraph were taken from Shermer, 160. Shermer's note for the source of this passage is: April 28, 1869; in Lyell, v. ii, 442.

15. Quoted from Shermer, 158. His note reads: "Ibid., 391–392." [This refers to WB 17, I. This is Wallace's autobiography, *My Life*, published in 1905.]

16. Shermer (2002), 160. Emphasis is in the original.

17. An excellent summary of Collins's views is found in the booklet "Does Evolution Explain Human Nature." It was published in 2010 by the John Templeton Foundation, and Collins's essay is found on pages 16–18 of the booklet.

18. T. Lombrozo, "Can science deliver the benefits of religion?" *Boston Review*, August 7, 2013.

19. S. K. Brem, M. Ranney, and J. Schindel, "Perceived consequences of evolution: College student perceive negative personal and social impact in evolutionary theory," *Science Education* 87 (2003): 181–206.

20. Robinson, *The Death of Adam*, 30–31.

21. Ibid.

22. R. Kimball, "John Calvin got a bad rap," *New York Times Review of Books*, February 7, 1999.

23. Robinson, *The Death of Adam*, 74–75.

24. R. Dawkins, *River Out of Eden*, 133.

25. The source for these remarks is the transcript of a YouTube video presentation by Neil deGrasse Tyson published on January 5, 2013, at: https://www.youtube.com/watch?v=ZkIEkejNF-8.

26. S. J. Gould, "Spin-doctoring Darwin," *Natural History* 104 (July 1995): 6–9.

27. H. Gee, *The Accidental Species*, 9.

28. Ibid., 12.

29. Ibid., 106.

30. J. Mascaro et al., "Testicular volume is inversely correlated with nurturing-related brain activity in human fathers," *Proceedings of the National Academy of Sciences* 110 (2013): 15746–15751.

31. This is the "Urban Initiative" of the Evolution Institute, based in Rochester, New York, and headed by evolutionary psychologist David Sloan Wilson.

32. The title of the conference was "The Two Cultures in the Twentieth Century," held in New York City at the NY Academy of Sciences on May 9, 2009.

33. Harris, *Free Will*, 47.

34. Letter to William Graham, July 3, 1881. Quoted from the online Darwin correspondence project: darwinproject.ac.uk/letter/entry-13230.

CHAPTER 2: Say It Ain't So

1. Any number of people have been suspected of perpetrating the Piltdown fraud, from the Jesuit paleontologist Pierre Teilhard de Chardin to Arthur Conan Doyle, the creator of the Sherlock Holmes mysteries. It now seems clear, however, that the person who actually found the fossils, Charles Dawson, carried out the fraud on his own. A brief, unsigned report of this appeared in the August 12, 2016, issue of *Science* magazine (vol. 353, p. 629). It summarized a much more detailed study of the hoax that had appeared in a journal of the British Royal Society: De Groote, I. et al., "New genetic and morphological evidence suggests a single hoaxer created 'Piltdown man,'" *Royal Society Open Science* 3 (2016): 160328.

2. A website describing these finds and continuing research at the Dmanisi site: http://www.dmanisi.ge/.

3. 145 centimeters is roughly 4 feet 9 inches.

4. 45 kilograms is equivalent to 99 pounds.

5. The details of this study were reported in D. Lordkipanidze et al., "A complete skull from Dmanisi, Georgia, and the evolutionary biology of early Homo," *Science* 342 (2013): 326–331.

6. This point of view was argued in a 2014 article that considered a "package" of traits unique to *Homo sapiens* and came to the conclusion that at least three distinct lineages of our genus existed at the same time as the Dmanisi fossils. See S. C. Antón et al., "Evolution of early *Homo*: An integrated biological perspective," *Science* 344 (2014): 1236828.

7. These assertions appeared in an ICR-sponsored web page authored by Brian Thomas and Frank Sherwin, first posted in 2013, and accessed January 13, 2016, at: http://www.icr.org/article/human-like-fossil-menagerie -stuns-scientists/.

8. B. Thomas, "New 'human' fossil borders on fraud," 2013. The article, posted on November 13, 2013, appeared at: http://www.icr.org/article/ new-human-fossil-borders-fraud/.

9. Miller, *Only a Theory*, 95. The discussion of creationist reactions to these fossils appears on pages 93–95 and is summarized in Figure 4.2.

10. Ibid., 95.

11. C. DeMiguel and M. Hennenberg, "Variation in hominid brain size: How much is due to method?" *Homo* 52 (2001): 3–58.

12. Historians of the field differ a bit as to when and how the neo-Darwinian synthesis emerged. But there seems to be a general consensus that it was developed in the late 1930s and early 1940s. Julian Huxley may have helped to coin the term with his 1942 book, *Evolution: The Modern Synthesis*. It is quite clear, however, that these updated versions of evolutionary theory were in place before DNA was generally understood to be the molecule of inheritance.

13. A recent study describing these roles is: C. Freyer and M. Renfree, "The mammalian yolk sac placenta," *Journal of Experimental Zoology* 312B(2009): 545–554. In their summary, these authors write: "The metabolic and biosynthetic functions of the yolk sac of the ancestral therian stem species, as well as hematopoiesis in the eutherian ancestor, appear to have been retained by the human yolk sac." So

there is no question that the yolk sac remains functional, if much reduced in size when compared to vertebrates that produce large, yolky eggs."

14. This is, of course, a gross oversimplification of the actual process of gene expression. For the sake of brevity and clarity in the service of general readers, I've skipped over such things as transcription factors, RNA polymerase, RNA splicing, RNA editing, and polyadenylation, as well as terms like *messenger RNA, transfer RNA*, and *ribosomal RNA*. The interested reader may, of course, consult a general biology textbook, at either the high school or college level, to find a more detailed discussion of gene expression and protein synthesis. My goal at this point is merely to give the nontechnical reader a sense of the relationship among genes, RNAs, and proteins that will help to explain the story of vitellogenin gene remnants in the human genome.

15. This work is described in: D. Brawand et al., "Loss of egg yolk genes in mammals and the origin of lactation and placentation," *PLoS Biology* 6 (2008): e63.

16. Gould, *The Panda's Thumb*, 28–29.

17. As it turns out, the story got even better as the researchers dug into their data. By comparing the loss of functional VIT genes across several genomes, they were able to estimate how long ago these sequences were converted to pseudogenes. They found that genes for the milk protein, casein, appeared just as the first of several VIT sequences became nonfunctional. Describing the significance of this, they explained that these data are compatible with a model in which lactation, the production of milk, appeared first in the common ancestor of all mammals. Then, as milk production increased, this allowed for the gradual reduction in the importance of yolk as a food source for the developing embryo. In terms of time, space, and evolutionary history, the various elements of the story fit together in a remarkable way.

18. Fairbanks, *Relics of Eden*.

19. Fairbanks, *Evolving: The Human Effect and Why It Matters*.

20. Processed pseudogenes are easy to recognize. When the DNA sequence of a typical gene is transcribed (copied into RNA), it contains regions known as *introns* (intervening sequences) that are cut out and removed before the RNA is actually used to guide the synthesis of a protein. When a RNA molecule that has been processed in this way is copied

back into DNA, the pseudogene that results is missing the intron sequences. So, a pseudogene that results from a DNA copying error generally contains those intron sequences, while a processed pseudogene does not.

21. Source for this estimate: D. J. Fairbanks et al., "NANOGP8: Evolution of a human-specific retro-oncogene," *G3* 2 (2012): 1447–1457.

22. H. A. Booth and P. W. Holland, "Eleven daughters of NANOG," *Genomics* 84 (2004): 229–238.

23. Centromeres are also the places where chromosomes remain briefly joined to each other after they have been duplicated prior to cell division. They are also the points where an attachment forms between each of the duplicate chromosomes and the fibers of mitotic spindle, which separate the duplicates during cell division.

24. This work was described in a 2005 paper: L. Hillier et al., "Generation and annotation of the DNA sequences of human chromosomes 2 and 4," *Nature* 434 (2005): 724–731.

CHAPTER 3: Chance and Wonder

1. These perplexing details of the life of Cain appear in the Book of Genesis, chapter 4, verses 13–17.

2. Written in Latin, the original title was *Prae-Adamitae*. One of the principal aims of his work, of course, was to clarify some of the stories found in the Book of Genesis. In Peyrère's view, the Genesis narrative presented Adam and Eve as the sole progenitors of the Jewish people. The existence of separately created pre-Adamite peoples not only solved certain mysteries in Genesis, such as the origin of Cain's wife, but also helped to explain the origins of aboriginal peoples such as the native American tribes, that did not seem to fit into the Genesis narratives.

3. Linnaeus's given name was actually Carl von Linné. However, the Latinized version of his name is generally attached to his works today, since they were published in Latin.

4. I have based a translation of Linnaeus's words here on a consensus excerpt found in Wikipedia, supplemented by my own very inexpert high school Latin. The original Latin text of this portion of Linnaeus's letter is: *Non placet, quod Hominem inter anthropomorpha collocaverim, sed*

*homo noscit se ipsum. Removeamus vocabula. Mihi perinde erit, quo no-
mine utamur. Sed quaero a Te et Toto orbe differentiam genericam inter
hominem et Simiam, quae ex principiis Historiae naturalis. Ego certissime
nullam novi. Utinam aliquis mihi unicam diceret!*

5. The Latin original is: *Si vocassem hominem simiam vel vice versa omnes
   in me conjecissem theologos. Debuissem forte ex lege artis.*

6. The Latin wording is: *Quam ampla sunt Tua Opera! Quam sapienter Ea
   fecisti! Quam plena est Terra possessione Tua!*

7. Huxley, *Evidence As to Man's Place in Nature*, 125.

8. The cartoon appeared on p. 776 of *Harper's Weekly*, August 19, 1871.
   To drive home his point, Nast positioned the gorilla standing in front
   of an office door for the Society for the Prevention of Cruelty to Animals.

9. The cartoon, which appeared on April 1, 1871, p. 130, was titled "A
   logical refutation of Mr. Darwin's theory."

10. *Hornet* magazine, March 22, 1871.

11. *Le Petite Lune*, August 1878.

12. Darwin, *The Descent of Man*, 186.

13. Darwin, *The Descent of Man*, 202. The discussion by Charles Lyell re-
    ferred to in this quotation is found in two places: Lyell, *Elements of
    Geology*, 583–585, and Lyell, *Antiquity of Man*, 145.

14. Ibid., 200.

15. T. Huxley, "On fossil remains of man," *Proceedings of the Royal Institu-
    tion of Great Britain* 3 (1862): 420–422.

16. The sketch is dated April 21, 1868, in Darwin's own hand.

17. Dubois himself chose the genus name *Pithecanthropus*, meaning "ape-
    man."

18. R. Dart, *"Australopithecus africanus*: The man-ape of South Africa,"
    Nature 115 (1925): 195–199.

19. Howell, *Early Man*.

20. Zallinger completed *The March of Progress* at the same time he was
    working on one of the largest scientific art projects ever conceived, *The
    Age of Reptiles*, a spectacular 110-foot mural in the Yale Peabody Mu-
    seum.

21. If my own memories of this time are still strong, it is for a good reason.
    In the summer of 1964 I was part of the Boy Scout pavilion, guid-
    ing visitors through the Scouting exhibits, acting as an honor guard

to visiting dignitaries, and conducting demonstrations of Scout craft and camping skills. Scouts from around the country were brought in as groups, put up at a nearby army base, and rotated through two-week tours in the Boy Scout Service Corps. We spent about four hours a day at our own exhibit, and had the rest of the day to tour the fair, enjoy the rides, the food, the international visitors, and the general ambience of the place. As a result, my friends and I toured many of the exhibits four, five, and even six times or more.

22. Sawyer and Deak, *The Last Human*.
23. Ibid., 18.
24. Ibid., 19.
25. Gould, *Bully for Brontosaurus*.
26. The diagram presented a linear version of horse evolution, showing a direct link from a single ancestral species to a single contemporary species (*Equus caballus*). It appeared in: W. Matthew, "The evolution of the horse: A record and its interpretation," *Quarterly Review of Biology* 1 (1926): 139–185.
27. B. MacFadden, "Fossil horses—Evidence for evolution," *Science* 307 (2005): 1728–1730.
28. Gould, *Bully for Brontosaurus*, 181.
29. Switek, *Written in Stone. Evolution, the Fossil Record, and Our Place in Nature*, 264.
30. The level of detail that a microscope allows us to "see" is typically presented in terms of its resolution, the smallest distance at which two objects can be resolved as being separate. For the transmission electron microscope, that is typically about 2 angstrom units, or 0.2 nanometers. A nanometer is 2×10^{-9} meters, which is quite close to the diameter of some atoms. However, for technical reasons, this level of resolution is very difficult to achieve with biological material. For thin slices of a cell, for example, the best resolution routinely achieved is about 10 times poorer, or 2 nanometers. While this usually does not allow us to see individual molecules, it does very clearly show a wealth of cells within that has allowed it to develop a very clear picture of the intricate complexity of living things.
31. Technically, we don't look at truly "living" cells in the transmission electron microscope, since living tissue usually has to be chemically stabilized, then dehydrated, and then cut into very, very thin slices to

endure the high vacuum inside the microscope. But over the years these preparation techniques have been refined to the point where we are confident that what we see does correspond to the structure of living cells.

32. F. deWaal, "Obviously, says the monkey," an essay in *Does Evolution Explain Human Nature* (West Conshohocken, PA: John Templeton Foundation, 2009).

33. Ibid.

34. Gould, *Wonderful Life: The Burgess Shale and the Nature of History.*

35. Ibid., 318.

36. Gee, *The Accidental Species*, 323.

37. H. Gee, "Brian Cox's human universe presents a fatally flawed view of evolution," *The Guardian*, October 14, 2014.

38. Gee, *The Accidental Species*, 189.

39. Ibid., 252.

40. E. Mayr, "The idea of teleology," *Journal of the History of Ideas*, 53 (1992): 117–135.

41. Gee, *The Accidental Species*, 69.

42. The complete quotation, from Dawkins's book *River Out of Eden*, is "The universe we observe has precisely the properties we should expect if there is, at bottom, no design, no purpose, no evil and no good, nothing but blind pitiless indifference."

CHAPTER 4: Explaining It All

1. Brownmiller, *Against Our Will: Men, Women and Rape.*

2. Ibid, 15. Emphasis shown in italics is in the original.

3. Ibid., 12.

4. See, for example: R. Bailey et al., "Rape behavior in blue-winged teal," *Auk* 95 (1978): 188–190, for specific descriptions of such behavior. Also, refer to M. Muller and R. Wrangham, *Sexual Coercion in Primates and Humans*, for descriptions of rape-like behavior in our closest relatives. Similar behaviors have also been reported in ducks, frogs, and even in certain species of worms and flies.

5. Anthropologist Donald Symons specifically argued against it in his landmark book, *The Evolution of Human Sexuality*. See p. 278 for a specific critique of Brownmiller's thesis.

6. R. Thornhill and C. T. Palmer, *A Natural History of Rape: Biological Bases of Sexual Coercion*.

7. Ibid., 12.

8. Ibid.

9. S. Begley, "Why do we rape, kill, and sleep around?" *Newsweek*, June 19, 2009.

10. I participated in this debate, which was titled "Resolved—That the Evolutionists Should Acknowledge Creation." The debate took place at Seton Hall University on December 4, 1997. The quotation presented here is taken from a video recording of the event made by PBS.

11. Thornhill and Barnes, *A Natural History of Rape*, 7.

12. It is worth noting that J. B. S. Haldane, *The Causes of Evolution*, suggested a solution to this problem as far back as 1932: "Insofar as it makes for the survival of one's descendants and near relations, altruistic behavior is a kind of Darwinian fitness, and may be expected to spread as a result of natural selection" (71). However, Haldane did not place this intuition on a theoretical basis, as Hamilton was to do some years later.

13. These papers are: W. D. Hamilton, "The genetical evolution of social behaviour. I," *J. Theor. Biol.* 7 (1964): 1–1, and W. D. Hamilton, "The genetical evolution of social behaviour. II," *J. Theor. Biol.* 7 (1964): 17–52.

14. Stevens was a remarkable scientist, and her story is too often ignored in histories of genetics. Determined to enter the new field of genetics, she taught high school for several years, then attended Stanford University, and subsequently earned her PhD at Bryn Mawr College. It was there that she began work with Thomas Hunt Morgan, the great pioneer of modern genetics. Among other achievements, Stevens discovered the Y chromosome, paving the way for our current understanding of sex-linked inheritance.

15. Wilson, *The Insect Societies*, 1.

16. Wilson, *Sociobiology: The New Synthesis*.

17. E. Allen et al., "Against 'Sociobiology,'" *New York Review of Books*, November 13, 1975.

18. Wilson described this incident in a book: Wilson, *Naturalist*, 348–350.

19. Wilson, *On Human Nature*.

20. Ibid., 169.

21. Ibid., 192.

22. A few have suggested that evolutionary psychology is nothing more than a "politically correct" version of sociobiology. There may be some truth to that charge, but it is more accurate to point out that the goals of evolutionary psychology are somewhat broader than sociobiology, since they encompass all of psychology and not just social behavior.

23. This quotation is taken from the home page of the Evolution Institute, a think tank organization founded by David Sloan Wilson. Accessed on January 29, 2015: https://evolution-institute.org/blog/welcome-to-the-evolutionary-blogosphere/.

24. For example, the sensory systems that drive this social behavior in *E. coli*, the common gut bacterium, have been described here: B. Ahmer, "Cell-to-cell signalling in *Escherichia coli* and *Salmonella enterica*," *Mol. Microbiol.* 52 (4) (2004): 933–945.

25. The technical symbol for the non-mutated copy of this gene is *fru*. You may be wondering how the *fruitless* mutation persists if such flies never produce offspring. The answer is that *fruitless* is recessive, meaning that it must be present in two copies to work its effects. Therefore, flies with just one copy of the mutation, whether male or female, can mate and produce offspring, many of which will carry two copies of the mutation.

26. L. Ryner et al., "Control of male sexual behavior and sexual orientation in *Drosophila* by the *fruitless* gene," *Cell* 87 (1996): 1079–1089.

27. C. Burr, "Homosexuality and biology," *The Atlantic*, June 1997.

28. See, for example, an article in the *New York Times* by science writer Nicholas Wade: "Mating game of fruit fly is traced to a single gene," December 13, 1996. Wade was willing to speculate about "the possibility that a related gene may operate in humans since the two species, despite their evolutionary distance, have equivalent versions of many important genes."

29. *Fruitless* codes for a so-called zinc finger protein that acts as a transcription factor, which can bind to other regions in the genome and activate specific sets of genes.

30. T. Shirango et al., "A double-switch system regulates male courtship behavior in male and female *Drosophila melanogaster*," *Nature Genetics* 38 (2006): 1435–1439.

31. For other examples, including an opening quotation from William Blake, see: R. J. Greenspan and H. A. Dierick, "'Am I not a fly like

thee?' From genes in fruit flies to behavior in human," *Human Molecular Genetics* 13 (2004): R267–R272.

32. A. Öhman and S. Mineka, "The malicious serpent: Snakes as a prototypical stimulus for an evolved module of fear," *Psychological Science* 12 (2003): 5–9.

33. J. C. Confer et al., "Evolutionary psychology: Controversies, questions, prospects, and limitations," *American Psychologist* 65 (2010): 221–126.

34. For a more detailed description of this work, see E. O. Wilson's book, *Consilience*, pp. 188–196. The original reference to the actual study is: A. P. Wolf, "Childhood association and sexual attraction: A further test of the Westermark hypothesis," *American Anthropologist* 72 (1970): 503–515.

35. Examples of such studies are: David Perrett et al., "Symmetry and human facial attractiveness," *Evolution & Human Behavior* 20 (1999): 295–307. K. Grammer and R. Thornhill, "Human (*Homo sapiens*) facial attractiveness and sexual selection: The role of symmetry and averageness," *Journal of Comparative Psychology* 108 (3) (1994): 233–42. B. C. Jones, et al., "Facial symmetry and judgements of apparent health support for a 'good genes' explanation of the attractiveness–symmetry relationship," *Evolution & Human Behavior* 22 (2001): 417–429.

36. B. Leach, "Shopping is 'throwback to days of cavewomen,'" *The Telegraph*, February 25, 2009.

37. D. Kruger and D. Byker, "Evolved foraging psychology underlies sex differences in shopping experiences and behaviors," *J. Social, Evolutionary, and Cultural Psychology* 3 (2009): 328–342.

38. Quotation taken from an online ABC News report by Lee Dye dated December 9, 2009. Accessed online January 6, 2016: http://abcnews.go.com/Technology/DyeHard/women-love-shop-men-dont-blame-evolution/story?id=9281875.

39. From Larry Moran's blog "Sandwalk." Accessed online January 6, 2016: http://sandwalk.blogspot.com/2009/03/shopping-is-throwback-to-days-of.html.

40. E. A. Smith et al., "Controversies in the evolutionary social science: A guide for the perplexed," *Trends in Ecology and Evolution* 16 (2001): 128–135.

41. As quoted by Sharon Begley in "Why do we rape, kill, and sleep around?" *Newsweek*, June 19, 2009.

42. For a summary of the case against Hauser, see: E. S. Reich, "Misconduct ruling is silent on intent," *Nature* 489 (2012): 189–190.

43. Hauser, *Moral Minds.*

44. The blogger was journalist Annalee Newitz. Her posting's URL is: http://io9.gizmodo.com/the-rise-of-the-evolutionary-psychology -douchebag-757550990.

45. Williams, *The Pony Fish's Glow*, 156–157.

46. M. Daly and M. I. Wilson, "An assessment of some proposed exceptions to the phenomenon of nepotistic discrimination against stepchildren," *Ann. Zool. Fennici* 38 (2001): 287–296.

47. Pinker, *How the Mind Works*, 207.

48. D. W Yu, and G. H. Shepard, "Is beauty in the eye of the beholder?" *Nature* 396 (1998): 321–322.

49. L. Germine et al., "Individual aesthetic preferences for faces are shaped mostly by environments, not genes," *Current Biology* 25 (2015): 2684–2689.

50. This monologue by actor John Cleese appears here on YouTube: https://www.youtube.com/watch?v=-M-vnmejwXo.

51. Wilson, *Consilience*, 286.

CHAPTER 5: The Mind of a Primate

1. From Darwin's letter to William Graham, dated July 3, 1881. Source: Darwin Correspondence Project, "Letter no. 13230," accessed on February 28, 2016.

2. Haldane, *Possible Worlds and Other Papers*, 286.

3. Shubin, *Your Inner Fish.*

4. Pinker, *How the Mind Works*, 21.

5. Zimmer, *Soul Made Flesh.*

6. Lewis, *The Weight of Glory and Other Addresses*, 139.

7. Haldane, *Possible Worlds*, 209.

8. Marcus, *Kluge: The Haphazard Construction of the Human Mind.*

9. F. de Waal, "Obviously, says the monkey." In *Does Evolution Explain Human Nature*, a booklet published by the Templeton Foundation, 2008. Accessed online at: http://www.templeton.org/evolution.

10. de Waal, *Are We Smart Enough to Know How Smart Animals Are?*

11. de Waal, "Obviously, says the monkey."

12. Darwin, *On the Origin of Species*, 421.

13. S. Gould and R. Lewontin, "The spandrels of San Marco and the Panglossian paradigm: A critique of the adaptationist programme," *Proc. R. Soc. Lond. B.* 205 (1979): 581–598.

14. M. Florio, et al., "Human-specific gene ARHGAP11B promotes basal progenitor amplification and neocortex expansion," *Science* 347(2015): 1465–1470.

15. H. Pontzer et al., "Metabolic acceleration and the evolution of human brain size and life history," *Nature* 533 (2016): 390–392. For an excellent summary of this research see: A. Gibbons, "Why humans are the high-energy apes," *Science* 352 (2016): 639.

16. My colleague at Brown, Phillip Lieberman, a student of Chomsky, has argued this point throughout his career. A brief summary of his ideas is given here: P. Lieberman, "Language did not spring forth 100,000 years ago," *PLoS Biology* 13(2) (2015): e1002064.

17. R. L. Buckner and F. M. Krienen, "The evolution of distributed association networks in the human brain," *Trends in Cognitive Sciences* 17 (2013): 648–665.

18. Ibid., 648.

19. C. Zimmer, "In the human brain, size really isn't everything," *New York Times*, December 31, 2013, p. D3.

20. An example is the 2014 movie *Transcendence*, starring Johnny Depp.

21. R. Epstein, "The Empty Brain," *Aeon*, May 18, 2016.

22. Jeff Shallit and I were among the six expert witnesses for the plaintiffs at the 2005 *Kitzmiller v. Dover* trial on evolution and intelligent design. Because a certain expert witness for the Dover School Board chose not to testify, Jeff's testimony was not needed in rebuttal, so he did not appear at the trial. But he nonetheless formed an important part of our team supporting the parents of Dover schoolchildren who were the victorious plaintiffs in the case.

23. Shallit's essay appeared online at: http://recursed.blogspot.com/2016/05/yes-your-brain-certainly-is-computer.html.

24. See, for example, technology writer Alex Knapp's article in *Forbes* magazine: "Why your brain isn't a computer," May 4, 2012.

25. K. D. Miller, "Will you ever be able to upload your brain?" *New York Times*, October 10, 2015, p. SR6.

26. G. Marcus, "Face it. Your brain is a computer," *New York Times*, June 28, 2015, p. SR12.

27. The candidate was former governor of Arkansas, Mike Huckabee. He was quoted in a transcript of a Republican primary debate published in the *New York Times* on June 5, 2007. Online at: http://www.nytimes .com/2007/06/05/us/politics/05cnd-transcript.html.

28. Marcus, *Kluge*, 176.

29. Wilson, *Consilience*, 108.

30. Robinson, *Absence of Mind*, 112.

CHAPTER 6: Consciousness

1. Chalmers used this metaphor in a March 2014 TED talk, "How do you explain consciousness?" archived online at: https://www.ted.com/talks/ david_chalmers_how_do_you_explain_consciousness.

2. G. Johnson, "Science of Consciousness conference is carnival of the mind," *New York Times*, May 17, 2016, p. D5.

3. R. Nagel, *Mind and Cosmos. Why the Materialist Neo-Darwinian Conception of Nature Is Almost Certainly False* (New York: Oxford University Press, 2012).

4. D. J. Chalmers, "Facing up to the problem of consciousness," *Journal of Consciousness Studies* 2 (3) (1995): 200–219.

5. A. Del Cul et al., "Brain dynamics underlying the nonlinear threshold for access to consciousness," *PLoS Biology* 5 (2007): e260.

6. Dehaene, *Consciousness and the Brain*, 134.

7. T. Nagel, "What is it like to be a bat?" *Philosophical Review* LXXXIII (October 1974): 435–450.

8. Ibid.

9. One way to understand the distinction between a scientific description of light and the experience of seeing it might be to consider the case of ultraviolet light. On the spectrum of electromagnetic radiation, ultraviolet (UV) light lies below the portion of that spectrum that is visible to us, so UV light is invisible to us. Despite the fact that we cannot see UV light, we can still produce it, detect its presence, and measure its effects on matter. In laboratory work, we routinely measure the absorption of UV light by solutions of proteins and nucleic acids. Therefore, we have a complete scientific understanding of the nature of UV light. Now,

let's imagine speaking with a person who can "see" in the ultraviolet in the same way we can see in the visible region of the spectrum. To them, UV light would surely be just another "color" they would experience like blue or green, and yet they would never be able to convey to us the exact experience of seeing objects in the ultraviolet "color." Seeing in the UV would change their experience of the world in a way that we cannot scientifically describe. Significantly, this is not a purely hypothetical example. Insects see quite well in the UV, and many flowers have pigmentation patterns that attract bees by reflecting distinctive patterns visible to their pollinators only in the UV.

10. Dickey, *The Eye-Beaters, Blood, Victory, Madness, Buckhead and Mercy.*

11. For a summary of this work, see: K. Heyman, "The map in the brain: Grid cells may help us navigate," *Science* 312 (2006): 680–681. A popularized version of this work appeared in: M.-B. Moser and E. I. Moser, "Where am I? Where am I going?" *Scientific American* 314 (1) (January 2016): 26–33.

12. A. G. Huth et al., "Natural speech reveals the semantic maps that tile human cerebral cortex," *Nature* 532 (2016): 453–458.

13. Along these lines, it might be worth noting that the neural correlates of the sensation of thirst have been well mapped over years of research. Using dogs as experimental subjects, investigators have located regions in the brain that are activated when the salt content of the bloodstream is artificially increased. With that knowledge, it has been possible to either suppress or to enhance the sensation of thirst in these animals by electrical stimulation of certain clusters of neurons. Therefore, in cellular and physical terms, we are very close to a complete understanding of how the sensation of thirst is produced. That's interesting, of course, and crucial to our understanding of how the brain produces sensation. However, nothing in this research captures or accounts for the actual feeling of being thirsty. Both are "hard" problems, but the inner, conscious feeling of thirst remains beyond experimental explanation even if we learn the circuitry that triggers it. That is the unapproachable part of the hard problem.

14. Nagel, *Mind and Cosmos*, 35.

15. Tallis, *Aping Mankind.*

16. Ibid., 119.

17. Nagel, *Mind and Cosmos*, 6.

18. H. A. Orr, "Awaiting a new Darwin," *New York Review of Books*, February 7, 2013.

19. Tallis, *Aping Mankind*, 171.

20. Ibid., 179.

21. Ibid., 241.

22. I suppose one could make an argument that consciousness is dependent upon the heart as well, since without a heart, there is no consciousness for the very simple reason that heart failure causes death. But this is not quite right. Consider the fact that many people whose hearts have been replaced by artificial mechanical pumps have survived for weeks and been quite conscious the whole time. No one would suggest that these pumps supplied consciousness to any of the patients in which they were implanted. Rather, they performed the mechanical, pumping function of the heart. The brain is different in this respect from every other part of the body in that it is directly linked to consciousness.

23. For the sake of clarity and brevity, I left out a parenthetical phrase that appears just before the last word in this quotation: " . . . no other material object—including most of the human nervous system and perhaps all of the nervous system of some lower animals—possesses." Tallis, *Aping Mankind*, 103.

24. Nagel, *Mind and Cosmos*, 36.

25. To be completely accurate, plants do this, too. Both plants and animals carry out cellular respiration, which releases carbon dioxide into the atmosphere. Only plants, of course (along with many types of microorganisms) can carry out photosynthesis, which removes carbon dioxide from the air.

26. A. B. Pippard, "The invincible ignorance of science," *Contemporary Physics* 29 (1988): 393–405.

27. C. McGinn, "All machine and no ghost," *New Statesman*, February 20, 2012. Accessed online July 1, 2016 at: http://www.newstatesman.com/ideas/2012/02/consciousness-mind-brain.

28. Tallis, *Neuromania*, 5.

29. L. Chang and D. Y. Tsao, "The code for facial identity in the primate brain," *Cell* 169 (2017): 1013–1028.

30. The source for this quotation is: N. Wade, "You look familiar. Now scientists know why," *New York Times*, June 6, 2017 p. D3.

31. Schrödinger, *What Is Life?*

32. Ibid., 68.
33. Lane, *Life Ascending: The Ten Great Inventions of Evolution*, 233.
34. Nagel, *Mind and Cosmos*, 105.
35. Tallis, *Aping Mankind*, 12.
36. Ibid., 349.
37. G. Strawson, "Consciousness isn't a mystery. It's matter," *New York Times*, May 16, 2016.
38. Lane, *Life Ascending*, 259.

CHAPTER 7: I, Robot

1. Wright, *The Moral Animal. Evolutionary Psychology and Everyday Life*, 350.
2. D. P. Barash, "Dennett and the Darwinizing of free will," *Human Nature Review* 3 (2003): 222–225.
3. W. Provine, "Evolution: Free will and punishment and meaning in life." Quoted from an abstract of Provine's keynote address, given at the second annual Darwin Day at the University of Tennessee, February 12, 1998. Online at: https://web.archive.org/web/20070829083051/http://eeb.bio.utk.edu/darwin/Archives/1998ProvineAbstract.htm.
4. S. Cave, "There's no such thing as free will," *The Atlantic*, June 2016, 69–74.
5. Ibid., 70.
6. This is from a notebook entry by Darwin dated September 6, 1838.
7. Darwin, *The Descent of Man*, 89.
8. Dennett, *Freedom Evolves*, 15.
9. Searle, *Freedom and Neurobiology: Reflections on Free Will, Language, and Political Power*, 37.
10. Boswell, *The Life of Samuel Johnson*, 169.
11. J. Locke, from a letter to Molyneux, January 20, 1693, in *The Correspondence of John Locke*, vol. IV (Oxford: Clarendon Press, 1979), 625.
12. Cottingham et al., *The Philosophical Writings of Descartes*, Volume I, 206.
13. Ibid., 99.
14. Ibid., 108.
15. The pineal gland is involved in circadian rhythms associated with sleep and wakefulness. One of its principal secretions is the hormone melatonin.

16. Cottingham et al., *The Philosophical Writings of Descartes*, Volume III, 206.
17. Wilson, *Consilience*, 99.
18. Pinker, *How the Mind Works*, 924.
19. From the transcript of a short video entitled "Steven Pinker on Free Will." Accessed online July 18, 2016, at: http://bigthink.com/videos/steven-pinker-on-free-will.
20. Ibid. Pinker video.
21. Harris, *Free Will*, 56.
22. Lucretius, *On the Nature of the Universe (De Rerum Natura)*, 44.
23. Dehaene, *Consciousness and the Brain*, 264–265.
24. Ibid., 265.
25. Harris, *Free Will*, 5.
26. These are the experiments of Benjamin Libet and others, which will be considered in depth a bit later in this chapter.
27. Harris, *Free Will*, 47.
28. Ibid., 19.
29. Ibid., 7.
30. Ibid., 43.
31. Ibid., 65.
32. Hawking, *A Briefer History of Time*, 17.
33. From the introduction to Laplace's *Essai philosophique sur les probabilités*, 1814.
34. P. W. Anderson, "More is different," *Science* 177 (1972): 393–396.
35. Ibid., 394.
36. Penrose, *The Emperor's New Mind*.
37. S. Hameroff, "How quantum brain biology can rescue conscious free will," *Frontiers in Integrative Neuroscience* 6 (93) (2012): 1–17.
38. J. R. Reimers et al., "The revised Penrose-Hameroff orchestrated objective-reduction proposal for human consciousness is not scientifically justified," *Physics of Life Reviews* 11 (2014): 101–103.
39. Tse, *The Neural Basis of Free Will*.
40. Ibid., 1.
41. J. Scholz, et al., "Training induces changes in white-matter architecture," *Nature Neuroscience* 12 (2009): 1370–1371.
42. A. A. Schlegel et al., "White matter structure changes as adults learn a second language," *J. Cognitive Neuroscience* 24 (2012): 1664–1670.

43. Ibid., 22.

44. P. U. Tse, "Free will unleashed," *New Scientist* 218 (2013): 28–29.

45. See, for example, pp. 8–9 in *Free Will*, by Sam Harris.

46. C. S. Soon et al., "Unconscious determinants of free decisions in the human brain," *Nature Neuroscience* 11 (2008): 543–545.

47. See Dennett's extensive discussion of Libet's experiments in *Freedom Evolves*, 227–242.

48. Dennett, *Freedom Evolves*, 241.

49. Ibid., 242.

50. See, for example: H. G. Jo et al., "Simultaneous EEG fluctuations determine the readiness potential: Is preconscious brain activation a preparation process to move?" *Experimental Brain Research* 231 (2013): 495–500. Also: A. G. Guggisberg, "Timing and awareness of movement decisions: Does consciousness really come too late?" *Frontiers in Human Neuroscience* 7 (2013): article 385.

51. D. P. Barash, "Dennett and the Darwinizing of free will" (a review of *Freedom Evolves*, by Daniel Dennett), *Human Nature Review* 3 (2003): 222–225.

52. Dennett, *Freedom Evolves*, 267.

53. R. F. Baumeister, "Do you really have free will?" *Slate*, September 25, 2013.

54. Ibid.

55. Carter, *Mapping the Mind*, 201.

CHAPTER 8: Center Stage

1. Robinson, *The Death of Adam*, 62.

2. Dawkins, *The Selfish Gene*, 4.

3. The lyrics to this song were printed in a June 2005 article in *Cincinnati* magazine, "In Genesis," 134.

4. Robinson, *The Death of Adam*, 74–75.

5. His original title was *Stammbaum des Menschens*, which might be translated more literally as *The Origin Tree of Humanity*.

6. According to Timothy Shanahan (*The Evolution of Darwinism*, p. 288), this was a note to himself that Darwin wrote in the margins of his own copy of Robert Chambers's book, *Vestiges of the Natural History of Creation*.

7. Bronowski, *The Ascent of Man*, 19.

8. Ibid., 412.

9. Gould, *Full House*, 29.

10. Gould, *Wonderful Life*, 288–289.

11. R. Wright, "The intelligence test: Stephen Jay Gould and the nature of evolution," *The New Republic*, January 29, 1990.

12. Gould, *Wonderful Life*, 291.

13. R. Wright, "The intelligence test: Stephen Jay Gould and the nature of evolution," *The New Republic*, January 29, 1990.

14. For a point-counterpoint on contingency and the Burgess shale, look to dueling articles written by Gould and Conway Morris for *Natural History* magazine: S. Conway Morris and S. J. Gould, "Showdown on the Burgess Shale," *Natural History* 107 (10) (1998): 48–55.

15. Conway Morris, *The Runes of Evolution*, 7.

16. C. B. Albertin et al., "The octopus genome and the evolution of cephalopod neural and morphological novelties," *Nature* 524 (2016): 220–224.

17. Conway Morris, *The Runes of Evolution*, 20.

18. Crick, *The Astonishing Hypothesis*, 3.

19. R. L. Buckner and F. M. Krienen, "The evolution of distributed association networks in the human brain," Trends in Cognitive Sciences 17 (2013): 648–665.

20. Lewis, *Mere Christianity*, 60.

21. Robinson, *Absence of Mind*, 112.

22. Wilson, *On Human Nature*, 156.

23. P. Bloom, "The war on reason," *The Atlantic*, March 2013.

24. Hawking, *A Brief History of Time*, 193.

25. Ruse, *Beyond Mechanism*, 417. It should be noted that Ruse continued with a bow to the father of genetics by adding, "aided a little by Gregor Mendel."

26. This is the title of a 1975 article in *American Biology Teacher*, by Theodosius Dobzhansky.

27. Wilson's exact quote from the home page of the Evolution Institute, which he heads, as noted in chapter 1 is, "Historians will look back upon the twenty-first century as a time when the theory of evolution, confined largely to the biological sciences during the twentieth century, expanded to include all human-related knowledge. As we approach the 1/6th mark of the twenty-first century, this intellectual

revolution is already in full swing. A sizeable community of scientists, scholars, journalists, and their readers has become fully comfortable with the statement 'Nothing about X makes sense except in the light of evolution,' where X can equal anthropology, art, culture, economics, history, politics, psychology, religion, and sociology, in addition to biology."

28. Dutton, *The Art Instinct*, 3.

29. Ibid., 101.

30. Ibid., 175.

31. A. Gottlieb, "The descent of taste," *New York Times*, January 29, 2009, p. BR12.

32. J. Lehrer, "Our inner artist," *Washington Post*, January 11, 2009.

33. Ibid.

34. M. Mattix, "Portrait of the artist as a caveman," *New Atlantis*, Winter/Spring 2013, 135.

35. This figure does not include the frozen polar land masses.

36. Robinson, *The Death of Adam*, 4.

37. E. O. Wilson, *The Meaning of Human Existence* (New York: W. W. Norton, 2014), 13.

38. Ibid, 13–14.

39. Genesis 2: 19–20.

40. Bronowski, *The Ascent of Man*, 437.

41. Nagel, *Mind & Cosmos*, 123.

42. Ibid., 124.

The Chromosome 2 Fusion Site

1. A human cell normally has 46 chromosomes. Sperm and egg cells, however, have just half that number (23). So, each of us inherit 23 chromosomes from our mother and 23 from our father, resulting in 23 chromosome pairs, for an aggregate total of 46 chromosomes.

2. Ijdo, J. W., et al. (1991). "Origin of human chromosome 2: An ancestral telomere-telomere fusion." Proceedings of the National Academy of Sciences 88: 9051-9055.

3. *Kitzmiller v. Dover* (2005) was a Federal trial in Pennsylvania prompted by the attempts of the Dover, PA, School Board to establish an "intelligent design" curriculum in the local high school.

4. This is one of the key points made in a 1997 paper from Carol Greider's laboratory on "knockout" mice, lacking the telomerase enzyme. After several generations, the number of telomere repeats becomes smaller and smaller, eventually resulting in end-to-end chromosome fusions of the very sort that produced human chromosome 2. M. A. Blasco et al., "Telomere shortening and tumor formation by mouse cells lacking telomerase RNA," *Cell* 91 (1997): 25–34.

5. This criticism was published in the in-house journal of the Institute for Creation Research, a well-known antievolution group. The reference is: J. P. Tomkins, "Alleged human chromosome 2 'fusion site' encodes an active DNA binding domain inside a complex and highly expressed gene—negating fusion," *Answers Research Journal* 6 (2013): 367–375.

6. V. Costa et al., "DDX11L: A novel transcript family emerging from human subtelomeric regions," *BMC Genomics* 10 (2009): 250.

7. Genetic nomenclature can get quite convoluted, and the WASH family of genes is no exception. In this case, the letters stand for "Wiskott-Aldrich Syndrome family Homolog."

8. P. Wang et al., "Case report: Potential speciation in humans involving Robertsonian translocations," *Biomedical Research* 24 (2013): 171-174.

9. P. Martinze-Castro et al., "Homozygosity for a Robertsonian translocation (13q14q) in three offspring of heterozygous parents," *Cytogenetics and Cellular Genetics* 38 (1984): 310–312.

Bibliography

Ahmer, Brian M. (2004). "Cell-to-cell signalling in *Escherichia coli* and *Salmonella enterica.*" *Mol. Microbiol.* 52: 933–945.

Albertin, Caroline B., et al. (2016). "The octopus genome and the evolution of cephalopod neural and morphological novelties." *Nature* 524: 220–224.

Allen, Elizabeth, et al. (1975). "Against 'sociobiology.'" *New York Review of Books*, November 13.

Anderson, Phillip W. (1972). "More is different." *Science* 177: 393–396.

Antón, Susan C., et al. (2014). "Evolution of early *Homo*: An integrated biological perspective." *Science* 344: 1236828.

Bailey, Robert O., et al. (1978). "Rape behavior in blue-winged teal." *Auk* 95: 188–190.

Barash, David P. (2003). "Dennett and the Darwinizing of free will." *Human Nature Review* 3: 222–225.

Baumeister, Roy F. (2013). Do you really have free will? *Slate*, September 25, 2013. Accessed online: http://www.slate.com/articles/health _and_science/science/2013/09/free_will_debate_what_does_free_will _mean_and_how_did_it_evolve.html.

Begley, Sharon. (2009). "Why do we rape, kill, and sleep around?" *Newsweek*, June 19.

Bloom, Paul. (2013) *The War on Reason. The Atlantic*, March.

Booth, H. Anne F., and Peter W. Holland. (2004). "Eleven daughters of NANOG." *Genomics* 84: 229–238.

Boswell, James. (1833). *The Life of Samuel Johnson*. New York: George Dearborn.

Brawand, David, et al. (2008). "Loss of egg yolk genes in mammals and the origin of lactation and placentation." *PLoS Biology* 6: e63.

Brem, S. K., M. Ranney, and J. Schindel. (2003). "Perceived consequences of evolution: College student perceive negative personal and social impact in evolutionary theory." *Science Education* 87: 181–206.

Bronowski, Jacob. (1973). *The Ascent of Man*. Boston: Little, Brown and Company.

Brownmiller, Susan. (1975). *Against Our Will: Men, Women and Rape*. New York: Simon and Schuster.

Buckner, Randy L. and Fenna M. Krienen. (2013). "The evolution of distributed association networks in the human brain." *Trends in Cognitive Science* 17: 649.

Burr, Chandler. (1997). "Homosexuality and biology." *The Atlantic*, June.

Carter, Rita. (1998). *Mapping the Mind*. San Francisco: University of California Press.

Cave, Stephen. (2016). "There's no such thing as free will." *The Atlantic*, pp. 69–74, June.

Chalmers, David J. (1995). "Facing up to the problem of consciousness." *Journal of Consciousness Studies* 2: 200–219.

Chapman, Matthew. (2008). *40 Days and 40 Nights: Darwin, Intelligent Design, God, Oxycontin, and Other Oddities on Trial in Pennsylvania.* New York: HarperCollins.

Confer, Jaime C., et al. (2010). "Evolutionary psychology: Controversies, questions, prospects, and limitations." *American Psychologist* 65: 221–226.

Conway Morris, Simon, and Stephen J. Gould. (1998) "Showdown on the Burgess Shale." *Natural History* 107: 48–55.

_____. *The Runes of Evolution.* (2015). West Conshohocken, PA: Templeton Press.

Cottingham, John, et al. (1985) *The Philosophical Writings of Descartes*, Volume I. Cambridge: Cambridge University Press.

_____. (1991) *The Philosophical Writings of Descartes*, Volume III. *The Correspondence*. Cambridge: Cambridge University Press.

Crick, Francis. (1994). *The Astonishing Hypothesis*. New York: Simon and Schuster.

Daly, Martin, and Margo I. Wilson. (2001). "An assessment of some proposed exceptions to the phenomenon of nepotistic discrimination against stepchildren." *Ann. Zool. Fennici* 38: 287–296.

Dart, Raymond. (1925). "*Australopithecus africanus*: The man-ape of South Africa." *Nature* 115: 195–199.

Darwin, Charles. (1859; 2009). *On the Origin of Species*. In E. O. Wilson, ed., *From So Simple a Beginning*. New York: Norton.

Darwin, Charles. (1871). *The Descent of Man and Selection in Relation to Sex*. London: John Murray.

Dawkins, Richard. (2016). *The Selfish Gene—40th Anniversary Edition*. New York: Oxford University Press.

_____. (1995). *River out of Eden*. New York: Basic Books.

Dehaene, Stanislas. (2014). *Consciousness and the Brain*. New York: Penguin Books.

Del Cul, Antoine, et al. (2007). "Brain dynamics underlying the nonlinear threshold for access to consciousness." *PLoS Biology* 5: e260.

De Groote, Isabelle, et al. (2016). "New genetic and morphological evidence suggests a single hoaxer created 'Piltdown man'." *Royal Society Open Science* 3: 160328, http://dx.doi.org/10.1098/rsos.160328.

Dennett, Daniel. (2003). *Freedom Evolves*. New York: Viking Press.

Descartes, Réne. (1644). *Principles of Philosophy*.

de Waal, Frans. (2009). "Obviously, says the monkey," an essay in *Does Evolution Explain Human Nature*. The John Templeton Foundation, West Conshohocken, PA.

_____. (2016). *Are We Smart Enough to Know How Smart Animals Are?* New York: Norton.

Dickey, James. (1971). *The Eye-Beaters, Blood, Victory, Madness, Buckhead and Mercy*. Atlanta: Hamish Hamilton Press.

Dutton, Denis. (2008). *The Art Instinct*. New York: Bloomsbury Press.

Epstein, Robert. (2016). "The empty brain." *Aeon*, May 18.

Fairbanks, Daniel J. (2007). *Relics of Eden. The Powerful Evidence of Evolution in Human DNA.* Amherst, New York: Prometheus Books.

_____. (2012). *Evolving: The Human Effect and Why It Matters.* Amherst, New York: Prometheus Books.

_____. et al. (2012). "NANOGP8: Evolution of a human-specific retro-oncogene." *G3 (Genes, Genomes, Genetics)* 2: 1447–1457.

Freyer, Claudia, and Marilyn B. Renfree. (2009). "The mammalian yolk sac placenta." *Journal of Experimental Zoology* 312B: 545–554.

Gee, Henry. (2013). *The Accidental Species.* Chicago: University of Chicago Press.

Germine, Laura, et al. (2015). "Individual aesthetic preferences for faces are shaped mostly by environments, not genes." *Current Biology* 25: 2684–2689.

Gibbons, Ann. (2016). "Why humans are the high-energy apes." *Science* 352: 639.

Gottlieb, Anthony. (2009). "The descent of taste." *New York Times*, January 29, p. BR12.

Gould, Stephen J., and Richard C. Lewontin. (1979). "The spandrels of San Marco and the Panglossian paradigm: A critique of the adaptationist programme." *Proc. R. Soc. Lond.* B. 205: 581–598.

Gould, Stephen J. (1980). *The Panda's Thumb.* New York: W. W. Norton.

_____. (1989). *Wonderful Life.* New York: Norton.

_____. (1991). *Bully for Brontosaurus.* New York: Norton.

_____. (1995). "Spin-doctoring Darwin." *Natural History* 104: 6–9.

_____. (1996). *Full House.* New York: Harmony Books.

Grammer, Karl, and Randy Thornhill. (1994). "Human (*Homo sapiens*) facial attractiveness and sexual selection: The role of symmetry and averageness." *Journal of Comparative Psychology* 108: 233–242.

Greenspan, Ralph J., and Herman A. Dierick. (2004). "'Am I not a fly like thee?': From genes in fruit flies to behavior in humans." *Human Molecular Genetics* 13: R267–R272.

Guggisberg, Adrian G. (2013). Timing and awareness of movement decisions: Does consciousness really come too late? *Frontiers in Human Neuroscience* 7: 1–11 (article 385).

Haldane, John Burdon Sanderson. (1932). *The Causes of Evolution.* London: Longmans, Greene, and Co.

_____. (1927). *Possible Worlds and Other Papers.* London: Chatto and Windus.

Hameroff, Stuart. (2012). "How quantum brain biology can rescue conscious free will." *Frontiers in Integrative Neuroscience* 6: 1–17.

Hamilton, William D. (1964). "The genetical evolution of social behaviour. I." *J. Theoretical Biology* 7: 1–16.

_____. (1964). "The genetical evolution of social behaviour. II." *J. Theoretical Biology* 7: 17–52.

Hauser, Marc R. (2006). *Moral Minds: How Nature Designed a Universal Sense of Right and Wrong.* New York: HarperCollins.

Harris, Sam. (2012). *Free Will.* New York: Free Press.

Hawking, Stephen. (1998). *A Brief History of Time.* New York: Bantam Dell.

_____. (2005). *A Briefer History of Time*. New York: Random House.

Heyman, Karen. (2006). "The map in the brain: Grid cells may help us navigate." *Science* 312: 680–681.

Hillier, LeDeana W., et al. (2005). "Generation and annotation of the DNA sequences of human chromosomes 2 and 4." *Nature* 434: 724–731.

Howell, F. Clark. (1965). *Early Man*. New York: Time-Life Books.

Humes, Edward. (2007). *Monkey Girl*. New York: HarperCollins.

Huth Alexander G., et al. (2016). "Natural speech reveals the semantic maps that tile human cerebral cortex." *Nature* 532: 453–458.

Huxley, Thomas H. (1862). "On fossil remains of man." *Proceedings of the Royal Institution of Great Britain* 3: 420–422.

_____. (1863). *Evidence As to Man's Place in Nature*. London: Williams and Norgate.

Jo, Han-Gue, et al. (2013). "Simultaneous EEG fluctuations determine the readiness potential: Is preconscious brain activation a preparation process to move?" *Experimental Brain Research* 231: 495–500.

Johnson, George. (2016). "Science of Consciousness conference is carnival of the mind." *New York Times*, May 17, D5.

Jones, B. C., et al. (2001). "Facial symmetry and judgements of apparent health support for a 'good genes' explanation of the attractiveness–symmetry relationship." *Evolution & Human Behavior* 22: 417–429.

Kimball, Roger. (1999). "John Calvin got a bad rap." *New York Times Book Review*, February 7.

Knapp, Alex. (2012). "Why your brain isn't a computer." *Forbes*, May 4.

Kruger, Daniel, and Dreyson Byker. (2009). "Evolved foraging psychology underlies sex differences in shopping experiences and behaviors." *J. Social, Evolutionary, and Cultural Psychology* 3: 328–342.

Lane, Nick. (2009). *Life Ascending: The Ten Great Inventions of Evolution*. New York: Norton.

Laplace, Pierre-Simon. (1814). *Essai philosophique sur les probabilities*. Courcier: Paris.

Leach, Ben. (2009). "Shopping is 'throwback to days of cavewomen.'" *The Telegraph*, February 25.

Lebo, Lauri. (2008). *The Devil in Dover.* New York: The New Press.

Lehrer, Jonah. (2009). "Our inner artist." *Washington Post*, January 11.

Lewis, Clive S. (1949). *The Weight of Glory and Other Addresses*. New York: HarperOne.

Lewis, C. S. (1952). *Mere Christianity*. New York: HarperCollins.

Lieberman, Phillip. (2015). "Language did not spring forth 100,000 years ago." *PLoS Biology* 13(2): e1002064.

Locke, John. (1979). *The Correspondence of John Locke* (Vol. IV). Oxford: Clarendon Press.

Lombrozo, Tania. (2013). "Can science deliver the benefits of religion?" *Boston Review*, August 7. Accessed online: https://bostonreview.net/arts-culture/can-science-deliver-benefits-religion.

Lordkipanidze, David, et al. (2013). "A complete skull from Dmanisi, Georgia, and the evolutionary biology of early *Homo*." *Science* 342: 326–331.

Lucretius. (1994). *On the Nature of the Universe (De Rerum Natura)* (translated by R. E. Latham). London: Penguin Books.

Lyell, Charles. (1838). *Elements of Geology*. London: John Murray.

_____. (1863). *Geological Evidence of the Antiquity of Man*. London: John Murray.

MacFadden, Bruce. (2005). "Fossil horses—evidence for evolution." *Science* 307: 1728–1730.

Marcus, Gary. (2008). *Kluge: The Haphazard Construction of the Human Mind*. New York: Houghton Mifflin.

_____. (2015). "Face it. Your brain is a computer." *New York Times*, June 28, p. SR12.

Mascaro, Jennifer, et al. (2013). "Testicular volume is inversely correlated with nurturing-related brain activity in human fathers." *Proceedings of the National Academy of Sciences* 110: 15746–15751.

Matthew, William D. (1926). "The evolution of the horse: A record and its interpretation." *Quarterly Review of Biology* 1: 139–185.

Mattix, Micah. (2013). "Portrait of the artist as a caveman." *The New Atlantis*, Winter/Spring.

McEwan, Ian. (2005). *Saturday*. New York: Random House.

McGinn, Colin. (2012). "All machine and no ghost." *New Statesman*, February 20.

Miller, Kenneth D. (2015). "Will you ever be able to upload your brain?" *New York Times*, October 10, p. SR6.

Miller, Kenneth R. (2007). *Only a Theory: Evolution and the Battle for America's Soul*. New York: Viking Press.

Moser, May-Britt, and Edvard Moser. (2016). "Where am I? Where am I going?" *Scientific American* 314 (1): 26–33.

Muller, Martin N., and Richard W. Wrangham. (2009). *Sexual Coercion in Primates and Humans*. Cambridge, MA: Harvard University Press.

Nagel, Thomas. (1974). "What is it like to be a bat?" *Philosophical Review* LXXXIII: 435–450.

_____. (2012). *Mind and Cosmos. Why the Materialist Neo-Darwinian Conception of Nature Is Almost Certainly False*. New York: Oxford University Press.

Öhman, Arne, and Susan Mineka. (2003). "The malicious serpent: Snakes as a prototypical stimulus for an evolved module of fear." *Psychological Science* 12: 5–9.

Orr, H. Allen. (2013). "Awaiting a new Darwin." *New York Review of Books*, February 7.

Penrose, Roger. (1989). *The Emperor's New Mind*. New York: Oxford University Press.

Perrett, David, et al. (1999). "Symmetry and human facial attractiveness." *Evolution & Human Behavior* 20: 295–307.

Pinker, Steven. (1997). *How the Mind Works*. New York: Norton.

_____. (2013). "Science is not your enemy." *The New Republic*, August 6.

Pippard, A. Brian. (1988). "The invincible ignorance of science." *Contemporary Physics* 29: 393–405.

Pontzer, Herman, et al. (2016). "Metabolic acceleration and the evolution of human brain size and life history." *Nature* 533: 390–392.

Reich, Eugenie Samuel (2012). "Misconduct ruling is silent on intent." *Nature* 489: 189–190.

Reimers, Jeffrey. R., et al. (2014). "The revised Penrose–Hameroff orchestrated objective-reduction proposal for human consciousness is not scientifically justified." *Physics of Life Reviews* 11: 101–103.

Robinson, Marilynne. (1998). *The Death of Adam*. New York: Houghton Mifflin.

_____. (2010). *Absence of Mind: The Dispelling of Inwardness from the Modern Myth of the Self*. New Haven, CT: Yale University Press.

Ruse, Michael. (2013). In *Beyond Mechanism*, Henning, B. G., and A. C. Scarfe, eds. New York: Lexington Books.

Ryner, Lisa C., et al. (1996). "Control of male sexual behavior and sexual orientation in *Drosophila* by the *fruitless* gene." *Cell* 87: 1079–1089.

Sawyer, G. J. and V. Deak. (2007). *The Last Human: A Guide to Twenty-Two Species of Extinct Humans*. New Haven: Yale University Press.

Schlegel, Alexander A., et al. (2012). "White matter structure changes as adults learn a second language." *J. Cognitive Neuroscience* 24: 1664–1670.

Scholz, Jan, et al. (2009). "Training induces changes in white-matter architecture." *Nature Neuroscience* 12: 1370–1371.

Schrödinger, E. (1944). *What is life? The physical aspect of the living cell*. Cambridge: Cambridge University Press.

Searle, John R. (2006). *Freedom and Neurobiology: Reflections on Free Will, Language, and Political Power*. New York: Columbia University Press.

Shanahan, Timothy. (2004). *The Evolution of Darwinism*. Cambridge: Cambridge University Press.

Shapiro, Adam R. (2013). *Trying Biology: The Scopes Trial, Textbooks, and the Antievolution Movement in American Schools*. Chicago: University of Chicago Press.

Shermer, Michael. (2002). *In Darwin's Shadow. The Life and Science of Alfred Russel Wallace*. New York: Oxford University Press.

Shirangi, Troy R., et al. (2006). "A double-switch system regulates male courtship behavior in male and female *Drosophila melanogaster.*" *Nature Genetics* 38: 1435–1439.

Shubin, Neil. (2007). *Your Inner Fish*. New York: Pantheon.

Slack, Gordie. (2007). *The Battle Over the Meaning of Everything. Evolution, Intelligent Design, and a School Board in Dover, PA*. San Francisco: John Wiley & Sons.

Smith, Eric A., et al. (2001). "Controversies in the evolutionary social science: A guide for the perplexed." *Trends in Ecology and Evolution* 16: 128–135.

Soon, Chun Siong, et al. (2008). "Unconscious determinants of free decisions in the human brain." *Nature Neuroscience* 11: 543–545.

Strawson, Galen. (2016). "Consciousness isn't a mystery. It's matter." *New York Times*, May 16.

Switek, Brian. (2010). *Written in stone. Evolution, the fossil record, and our place in nature*. New York: Bellevue Literary Press.

Symons, Donald. (1979). *The Evolution of Human Sexuality*. London: Oxford University Press.

Tallis, Raymond. (2011). *Aping Mankind*. New York: Routledge.

The John Templeton Foundation. (2010). *Does Evolution Explain Human Nature?* Published online: http://www.templeton.org/evolution/Essays/evolution_booklet.pdf.

Thomas, Brian, and Frank Sherwin. (2013). "Human-like fossil menagerie stuns scientists." Accessed January 3, 2016 at http://www.icr.org/article/human-like-fossil-menagerie-stuns-scientists/.

_____. (2013). "New 'human' fossil borders on fraud." Accessed January 3, 2016: http://www.icr.org/article/new-human-fossil-borders-fraud/.

Thornhill, Randy, and Craig T. Palmer. (1980). *A Natural History of Rape. Biological Bases of Sexual Coercion.* Cambridge, MA: MIT Press.

Tse, Peter U. (2013). *The Neural Basis of Free Will.* Cambridge, MA: MIT Press.

_____. (2013). "Free will unleashed." *New Scientist* 218: 28–29.

Wade, Nicholas. (1996). "Mating game of fruit fly is traced to a single gene." *New York Times*, December 13.

Williams, George C. (1997). *The Pony Fish's Glow.* New York: Harper Collins.

Wilson, Edward O. (1971). *The Insect Societies.* Cambridge, MA: Harvard University Press.

_____. (1975). *Sociobiology: The New Synthesis.* Cambridge, MA: Harvard University Press.

_____. (1978). *On Human Nature.* Cambridge, MA: Harvard University Press.

_____. (1994). *Naturalist.* Washington, DC: Island Press.

_____. (1998). *Consilience*. New York: Alfred A. Knopf.

_____. (2014). *The Meaning of Human Existence*. New York: W. W. Norton.

Wright, Robert. (1990). "A review of *Wonderful Life: The Burgess Shale and the Nature of History* by Stephen Jay Gould." *The New Republic*, January 29.

_____. (1995). *The Moral Animal. Evolutionary Psychology and Everyday Life*. New York: Vintage Books.

Wolf, Arthur P. (1970). "Childhood association and sexual attraction: A further test of the Westermark hypothesis." *American Anthropologist* 72: 503–515.

Yu, Douglas W., and Glenn H. Shepard. (1998). "Is beauty in the eye of the beholder?" *Nature* 396: 321–322.

Yunis, Jorge J., and Om Prakash. (1982). "The origin of man: A chromosomal pictorial legacy." *Science* 215: 1525–1530.

Zimmer, Carl. (2004). *Soul Made Flesh*. New York: Free Press.

_____. (2013). "In the human brain, size really isn't everything." *New York Times*, December 31, p. D3.

Acknowledgments

I am deeply grateful for the help, encouragement, and support I received from many individuals during the preparation of this book. These include Nick Matzke, Eugenie Scott, Rick Potts, and my Brown University colleagues Phillip Liebermann, William Fairbrother, Sohini Ramachandran, and the late Michael McKeown. I thank Randy Buckner, David Lordkipanidze, David Hillis, and David Fairbanks for permission to use illustrations from their work. I especially appreciate the kindness of Thomas Nagel in spending time with me to discuss some of the critical issues surrounding the evolution of consciousness. I must also acknowledge the great debt of inspiration I owe to authors such as Stephen Jay Gould, Jacob Bronowski, and E. O. Wilson. To me, their willingness to grapple with the great questions of human existence serves as a model for how science can inform and enrich our understanding of the world.

I thank my agent, Barney Karpfinger, who helped to shape and focus this manuscript over many years and through more than a few false starts, as well as my editor, Priscilla Painton. Their patience was exemplary and their support essential to seeing this book through to completion. For nearly forty years I have been privileged to teach and do research at Brown University, and in the process have been enriched by wonderful colleagues, creative students, and a profoundly supportive academic environment. Finally, I thank my family and especially my wife, Jody, for their forbearance, understanding, and love. Such gifts are precious without price and forever beyond my understanding.

Index

NOTE: *Italics* page numbers refer to figures.

Hunter, George, 11
Huxley, Thomas Henry, 55–56, 58, 64–65

incest, 100–101
infanticide, 107–8, 111
Institute for Creation Research (ICR), 34, 35, 236
intelligence, 21, 23–24, 79, 131–37, 162, 202, 209, 210–11, 212, 213, 214, 229
  See also brain; mind; thought
intelligent design, 2
intentionality, 227

Java: fossils in, 30, 59
jealousy, 100
John Deere Corporation, 24, 221
Johnson, George, 151
Johnson, Samuel, 177

Kafka, Franz, 155, 156
Kant, Immanuel, 220
Keith, Arthur, 59
kin selection, 88–92
Kinsley, Michael, 84
Kitzmiller v. Dover (Pennsylvania, 2005), 2–3, 236
Kluge (Marcus), 139
Koch, Christof, 154
Krienen, Fenna M., *36*, 136
Kruger, Daniel, 102–3

labor, division of, 93, 102–8, 111, 114
Landry, Sarah, 92
Lane, Nick, 172, 174
language, 21, 74, 135–36, 139, 159, 192, 202
Laplace, Pierre-Simon, 186
Larkin, Philip, 8
Lawrence, Jerome, 12
Leakey, Louis, 30
learning, 77, 99, 114
Leary, Timothy, 126
legs, 33–34
  See also bipedality

Lewis, C.S., 125, 216
Lewontin, Richard C., 93, 133–34
Libet, Benjamin, 194, 195
life
  ascent of, 205–9
  and consciousness, 165–67
  "deep structure" to, 213–14
  definition of, 146
  disagreements about origin of, 13
  as experiment, 208
  nature of, 165–67
  as physical process, 196
  properties of, 167
  ultimate potential of, 214
Linnaeus, Carolus, 53–55, 56, 67
literature, 9, 24.
  See also culture/society
lithium, 126
Locke, John, 177–78
Lombrozo, Tania, 17
Lordkipanidze, David, 33, 34, *37*
Lorenz, Konrad, 208
luck, 71–74, 75, 202
Lucretius, 182, 199
Luther, Martin, 28
Lyell, Sir C., 57

MacEnery, John, 53
MacFadden, Bruce, 65
MacLean, Paul, 130–31
Marcus, Gary, 130, 139, 143, 145
Marsh, O.C., 64–65
mathematics, 2, 4, 97, 112–13, 173, 202, 207, 224
mating, 106–7, 222.
  See also reproduction; sex
The Matrix (movie), 116–17
matter/materialism
  brain as, 120–21, 123–26, 146, 216–17
  and consciousness, 161, 168–69, 172, 174
  and free will, 186, 187
  and God, 216

About the Author

Kenneth R. Miller is professor of biology at Brown University where he teaches courses in cell biology and general biology. His scientific papers and reviews have appeared in leading journals including *Science, Cell, Nature,* and *Scientific American.* Miller is coauthor, with Joseph S. Levine, of high school and college biology textbooks used by millions of students nationwide. His honors include the AAAS Award for Public Engagement with Science, the Stephen Jay Gould Prize from the Society for the Study of Evolution, the Gregor Mendel Medal from Villanova University and the Laetare Medal from Notre Dame University.